ENVIRONMENTAL SCIENCE, ENGINEERING AND TECHNOLOGY

ANAEROBIC DIGESTION

TYPES, PROCESSES AND ENVIRONMENTAL IMPACT

Environmental Science, Engineering and Technology

Additional books in this series can be found on Nova's website under the Series tab.

Additional e-books in this series can be found on Nova's website under the e-book tab.

Waste and Waste Management

Additional books in this series can be found on Nova's website under the Series tab.

Additional e-books in this series can be found on Nova's website under the e-book tab.

ENVIRONMENTAL SCIENCE, ENGINEERING AND TECHNOLOGY

ANAEROBIC DIGESTION

TYPES, PROCESSES AND ENVIRONMENTAL IMPACT

ANA TORALES
EDITOR

nova publishers
New York

For permission to use material from this book please contact us:
Telephone 631-231-7269; Fax 631-231-8175
Web Site: http://www.novapublishers.com

Additional color graphics may be available in the e-book version of this book.

Library of Congress Cataloging-in-Publication Data

Anaerobic digestion (2013)
Anaerobic digestion : types, processes, and environmental impact / editor, Ana Torales.
pages cm
Includes bibliographical references and index.
ISBN 978-1-62808-884-7 (hardcover)
1. Sewage--Purification--Anaerobic treatment. I. Torales, Ana, editor of compilation. II. Title.
TD756.45.A54 2013
628.3'54--dc23
2013031605

Published by Nova Science Publishers, Inc. † New York

CONTENTS

PREFACE

Anaerobic digestion is used for industrial or domestic purposes to manage waste and/or to produce fuels. Anaerobic digestion are the group of processes by which microorganisms break down biodegradable material in the absence of oxygen. In this book, the authors present current research in the study of the types, processes and environmental impact of anaerobic digestion. Topics include the development and application of anaerobic digestion to treat husbandry and industrial wastewater in the Mekong Delta of Vietnam; anaerobic digestion of agricultural waste; evaluation of decentralized anaerobic digestion systems for wastewater treatment, reuse and nutrient recycling to alleviate water and food stress; effects of substrate-to-inoculum ratio on anaerobic digestion of vegetable wastes; anaerobic digestion sustainability effects on plant-wide performance and automation in centralized wastewater treatment; and anaerobic co-digestion of cheese whey and the screened liquid fraction of dairy manure.

Chapter 1 – This chapter analyses the development and the state of the art of anaerobic digester models to treat waste and wastewater discharged by both farming activities and industrial enterprises in the Mekong Delta of Vietnam. More than 50% of the national agricultural production is produced annually in the Mekong Delta, and anaerobic digester models are used to treat waste and wastewater in the region. In the agricultural sector, biogas plants play a key role in VACB integrated farming system to minimize the sanitation risks from husbandry at the suburban and rural areas. In the industrial sector, the anaerobic process is applied in the form of UASB reactor to treat wastewater from fish processing and beverage production enterprises. The chapter first gives a comprehensive introduction to the development of the key anaerobic digester models. Then it reviews the technical situation of each type of anaerobic models and examines its strengths and weaknesses. We find that although anaerobic digestion technology has been researched and applied there in the past 30 years, its development is still limited and need further studies to improve the anaerobic digester models currently applied in the region.

Chapter 2 – Agricultural waste represents one of the largest anthropogenic sources of methane present in the atmosphere. Anaerobic digesters can turn these wastes into biogas as renewable energy, containing at least 50% of methane, and the solid residues into fertilizers rich in nutrients. Anaerobic digestion has a long history (10th century BC biogas used to heat water in Assyria) and many rustic digesters are implemented all around the world in farms. Unfortunately the lack of understanding and mistrust resulted in numerous failures, e.g. China and India have reported 50% failure rate and closure, 80% for US farm digesters. So, the knowledge upon anaerobic digestion limitations and reasons for failures is important; design and engineering deficiencies, complex digestion processes and inappropriate equipment installed… There is no standard designed digester for different reasons: in one hand, the situations are very diverse in operating conditions and in the other one, there remain many research issues still incomplete concerning microbiology, microbial ecology, metabolic pathways, modelling, technology… First, this chapter will be a recall of basic principles of anaerobic digestion applied to this kind of complex substrate. A review of the main types of agricultural waste digesters used around the world, from the simplest to the most sophisticated will be made by presenting advantages and disadvantages of each including environmental impact. It will also compile state-of-the-art information related to new developments in anaerobic technology applied to agricultural waste treatment. But not only high technology could help in a correct control of the installation, and thus, a higher biogas production rate, also a balanced ration is required, and some simple recipes could help. Another aspect will be studied, namely the wishes of the manager of the facility. Indeed, depending on whether one wishes above all to produce more biogas (i.e. energy, therefore financial income), essentially to treat its effluents or produce a digestate agronomically interesting, driving of the installation and operation applied may differ.

Chapter 3 – Food shortage and drought is a growing concern in developing countries requiring economic reuse and nutrients recycling. Anaerobic digestion can be economically applied to treat domestic wastewater from rural and small communities. Upflow Anaerobic Sludge Blanket (UASB) is proven low-cost technology that can be sustainably built at small scale and locally operated for decentralized treatment and reuse. This book chapter evaluates typical decentralized UASB implementation in integration with crop production. The anaerobic process was evaluated for nutrient recovery, effluent reuse in drip irrigation system and for biogas production. N, P and K release in the effluent waters were evaluated as alternative fertilizer for tomato production in green house using drip irrigation. The UASB was followed by Rotated Biological Contactors (RBC), solids separation and filtration post treatment units to control chemical oxygen demand (COD) and suspended solids (SS) in the irrigation system. Biogas production was evaluated as a renewable source for energy generation and heat supply on a household scale. The health and environmental impacts were evaluated in the irrigation experiment according to the World Health Organization (WHO) and the Food Agriculture organization (FAO) guidelines.

The characteristics of the UASB effluent after post-treatment were: EC 3.0 dS/m, pH 7.7, BOD 303 mg/l, NO_3 5.1 mg/l, NH_4 54.7 mg/l, P 18.0 mg/l, K 36.8 mg/l, E. coli 31.7E+5 MPN/100ml and T. coliform 68.7E+5 MPN/100ml. These characteristics sufficed plant nutrient needs and resulted in 12% higher tomato crop yields compared to fertigation with inorganic fertilizer. However, salinity level was severe according to FAO. Although effluent counts of pathogenic indicators were not acceptable according to WHO, the counts were significantly reduced after irrigation as indicated by the analytical results of Escherichia coli and Total coliform in soil (<1.2E+3) and in plant (<1 MPN/100 ml).

Chapter 4 – The effects of substrate to inoculum ratio (SIR) on the anaerobic digestion of vegetable wastes slurries were investigated using biochemical methane potential (BMP) assays. Cow dung was used as inoculum for the BMP assays. The vegetable wastes consisted of a mixture cabbage tops, cauliflower leaves, carrots, beetroots, tomatoes and potatoes which was ground to form a slurry and characterized to yield a total solids (TS) of 6.40%, volatile solids (VS) of 76.92%, chemical oxygen demand (COD) of 11870.50mg/L, soluble COD (sCOD) of 3600.00 mg/L and a pH of 6.00. The different SIR was 1:1 and 3:1. 2 other set-up was done for 100% inoculum and 100% substrate. Water was added so as to have a TS of 8%. Digestion proceeded in a 2L reactor with a headspace of 500mL kept for storage of biogas. The TS, VS, pH, COD, sCOD, volatile fatty acids (VFA) were monitored for 34 days and the biogas production rates were monitored over a period of 60 days to assess the effects of SIR on the anaerobic digestion process chemistry. From the experiments, it was found that the pH varied from 5.72 to 9.05, VFA from 0 meq/L to 336.30 meq/L, COD from 3597.12mg/L to 18000mg/L and sCOD from 719.42mg/L to 5000mg/L. Highest VS reduction was achieved by reactor B (SIR 1:1). Volume of biogas produced for each BMP was as follows: 21200 ml for SIR 1:1, 18500 ml for SIR 3:1, for 100% inoculum we noted 1150 ml and for 100% substrate is 13050 ml of biogas. Results have shown that after 30 days, the BMP assay consisting of 50% inoculums yielded the highest cumulative production of biogas. However in the long run (after 60 days), an SIR ratio of 3:1 resulted in the highest biogas production rate.

Chapter 5 – In the absence of globally unified water quality regulations and with the growing sustainability concerns in the water sector, plant-wide modeling is a powerful tool to assist the selection of wastewater treatment technologies, optimize their operations and benchmark their sustainability. This book chapter illustrates these modeling benefits through advanced plant-wide modeling case study to evaluate the plant upgrade with anaerobic digestion (AD) and its effect on plant sustainability and control. We further discuss consequences of supernatant return after AD to the upstream Biological Nitrogen Removal (BNR) and possible technological solutions to avoid the nitrogen overload improving the treatment system sustainability. The International Water Association Benchmark Simulation Model (BSM) was upgraded with Green House Gas models for AD and BNR. Accordingly, the dynamic interactions and the impacts of the AD upgrade on the BNR system could be evaluated. The main interactions were the increased easily degradable substrate and ammonium load to the BNR due to anaerobic degradation of the sludge. This combined effect improved denitrification but increased nitrous oxide emissions and nitrate in the effluent. Aeration control reduced nitrate and nitrite concentrations and nitrous oxide emissions. However nitrous oxide emissions were still higher compared to the plant without AD. The water quality impacts remained within the regulated limits for the three treatment scenarios.

AD and aeration control upgrades increased the sustainability of the plant by reducing the sludge production, operational energy, and eutrophication impact by 50%, 70%, and 40%, respectively. The plant carbon footprint of WWT was increased with the AD upgrade due to increased nitrogen load from recycled supernatant and increased N_2O emissions. Reduction of the footprint depends on the post application of the AD effluents. In case of sludge disposal to landfill, AD would reduce the carbon footprint within the boundary of treatment and disposal by 15-20% due to reductions of sludge volumes.

Chapter 6 – Anaerobic co-digestion of animal manures and wastes or by-products from the food industry has become a key method for both waste treatment and the production of renewable fuels. In co-digestion processes, the proportion of substrates plays an important role due to synergies amongst them. However, the proportions employed must be carefully studied to avoid inhibitions due to toxic compounds or acidification. In this work, anaerobic co-digestion of cheese whey and screened liquid fraction of dairy manure has been investigated under batch mode operation. Six cheese whey proportions were tested, from 10 to 60%. Dairy manure and cheese whey resulted to be a good match for anaerobic co-digestion. Co-digestion of both substrates permitted to digest a difficult substrate such as the whey without alkalinity addition. However, above 20% cheese whey proportion, methane yield was lower than expected from methane potentials of both substrates separately. The best results in terms of methane conversion were obtained at 20% whey proportion. Under this condition methane yield of the mixture was 46.4% higher than that of the liquid manure, and cheese whey was completely degraded with an estimate methane yield of 18.4 L CH_4 kg^{-1} whey. On the contrary, with 60% cheese whey proportion, the methane yield was 82.7% higher than that of the liquid manure but the estimate methane yield of the whey was reduced to 13.2 L CH_4 kg^{-1}. Alkalinity of dairy manure was enough to develop the anaerobic co-digestion process. However, above 20% cheese whey proportion, methane yields were lower than expected.

In: Anaerobic Digestion
Editor: Ana Torales
ISBN: 978-1-62808-884-7

Chapter 1

Development and Application of Anaerobic Digestion to Treat Husbandry and Industrial Wastewater in the Mekong Delta of Vietnam

Nguyen Vo Chau Ngan and Le Hoang Viet
Department of Environmental Engineering
College of Environment and Natural Resources
Cantho University, Vietnam

Abstract

This chapter analyses the development and the state of the art of anaerobic digester models to treat waste and wastewater discharged by both farming activities and industrial enterprises in the Mekong Delta of Vietnam. More than 50% of the national agricultural production is produced annually in the Mekong Delta, and anaerobic digester models are used to treat waste and wastewater in the region. In the agricultural sector, biogas plants play a key role in VACB integrated farming system to minimize the sanitation risks from husbandry at the suburban and rural areas. In the industrial sector, the anaerobic process is applied in the form of UASB reactor to treat wastewater from fish processing and beverage production enterprises. The chapter first gives a comprehensive introduction to the development of the key anaerobic digester models. Then it reviews the technical situation of each type of anaerobic models and examines its strengths and weaknesses. We find that although anaerobic digestion technology has been researched and applied there in the past 30 years, its development is still limited and need further studies to improve the anaerobic digester models currently applied in the region.

Keywords: Anaerobic technology, husbandry waste treatment, the Mekong Delta

Introduction

According to Do (2005), Nguyen and Nguyen (2005), researches on anaerobic technology started from decade '60 of the last century in the Mekong Delta (MD). However, due to the massive imported liquefied gas and inorganic fertilizers at that time, the research findings could not be applied in reality. After the country re-union, biogas digester attracted the attention from the Vietnamese government that is a lever to put biogas gradually into practice. A research project on renewable energy (code 52C) took priority in the two periods of five-year national plans (from 1981 to 1985 and from 1986 to 1990). These research programs, leaded by Energy Science and Technology Research Institute, aimed to develop biogas technology.

In the MD, Ho Chi Minh City University of Technology and Cantho University were two local research partners involved in 52C project. There was large amount of household scale biogas plant installed in the MD during this phase, due to the favorable socio-economic and climate conditions. At the end of 1980s, some demonstration anaerobic digesters were built in Can Tho city - the biggest city in the MD - to treat husbandry's waste (household scale) and hospital's wastewater (at 30-4 Hospital). Then anaerobic technology spread to other provinces with around 500 units of household biogas plant was built in the MD. At this time there is no recorded on application of anaerobic digester to treat industrial wastewater in the MD.

From 1990 to 1993, anaerobic technology research and promotion activities were not involved in any nation planning. Without support from government, there are no remarkable events concerning to anaerobic technology in the MD and the whole of nation as well. Only since 1993, the anaerobic digester had strong growing in framework of environmental sanitation, agriculture cultivation and rural development projects. From these projects, the concept of an integrated farming system including of V (garden), A (fish pond), C (animal cage) and B (biogas plant) were developed and accepted. Many biogas models were developed and introduced to farmers which could apply as a component of VACB integrated farming system. In the MD, the most popular biogas models are PE digester, TG-BP, KT2, VACVINA, EQ1 and EQ2, composite, etc.

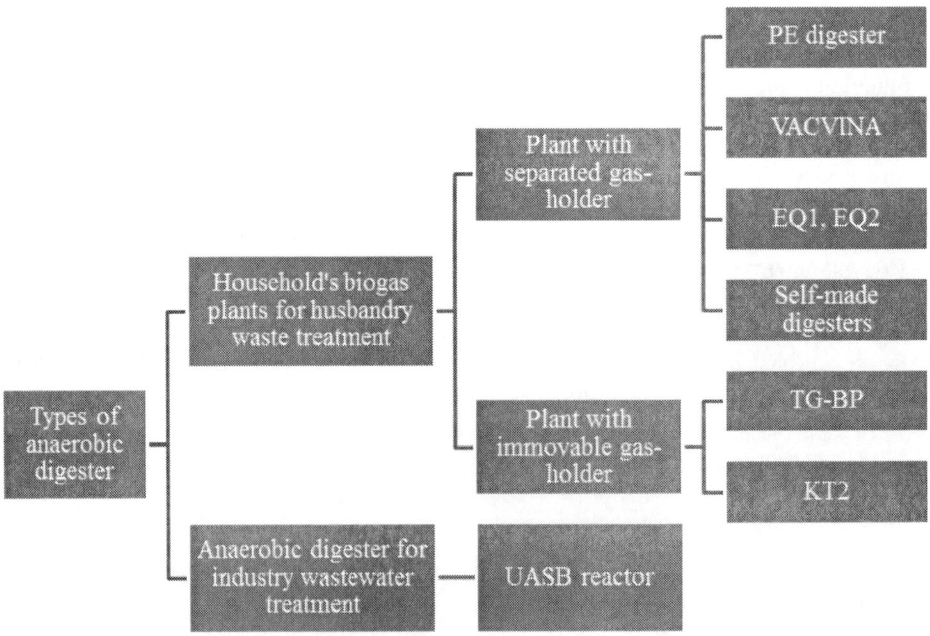

Figure 1. Types of anaerobic digesters currently used in the MD.

In recent years, the vast industrialization process in the MD causes of pollutions from industrial wastewater. To deal with this situation, anaerobic technology was introduced as a suitable way to treat for industrial wastewater. However the attractiveness of biogas digester for industrial scale applications has been limited because of its slow rate and process instability. So that, only Upflow Anaerobic Sludge Blanket (UASB) reactor was applied to treat wastewater from food processing and brewery production factories.

1. APPLYING BIOGAS TECHNOLOGY IN HUSBANDRY SECTOR

1.1. Overview of Biogas Digester at Household Scale

According to Nguyen (1989), the first biogas plant in the MD was introduced by Dr. Hoang Quynh in Can Tho province but the time was unknown. This plant was in parallel-piped shape and used water layer to keep constant gas pressure. During the time of 1985 and 1986, the Renewable Energy Center (REC) - Cantho University - developed some types of biogas digester such as rubber tube biogas digester, concrete biogas digester with rubber sheet cover.

In 1987, REC developed the CT1 biogas plant. The CT1 model is an upgraded model of floating drum model but with a fixed gas-holder made of ferro-cement. At that time, it was an innovated biogas plant. By this improvement, the CT1 model is airtight, high corrosion resistance, low cost and long life-time.

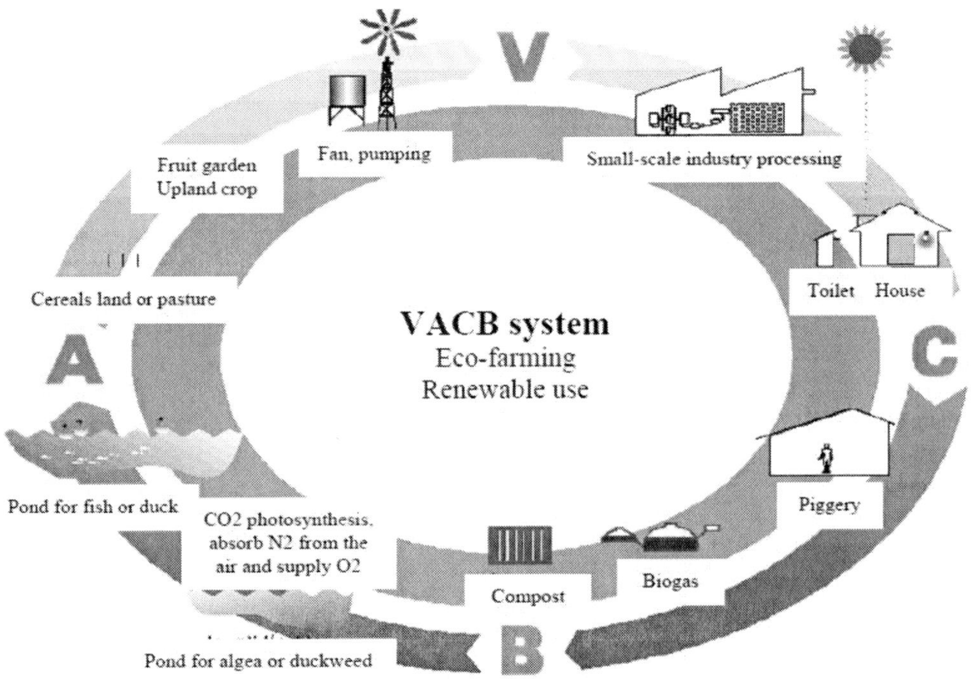

Source: REC, 1996.

Figure 2. The VACB integrated farming system.

The life-time of the CT1 plant is estimated at more than 10 years. By 1995, more than one hundred units were already constructed but the REC stopped manufacturing this biogas model (Do, 1989). Later on, due to its disadvantages in transportation of pre-casting components and the availability of some alternative biogas models, this plant is not considered attractive anymore.

In 1992, a project funded by SIDA and FAO studying strategies for sustainable use of local resources in livestock-based farming systems introduced the polyethylene (PE) digester in the South of Vietnam.

At the same time, REC introduced a new model to the farmers in the MD namely TG-BP biogas plant which developed by the scientists of Thailand and Germany within a GTZ project. The TG-BP plant integrated into a VAC farming system (V - garden, A - fishpond, C - animal cage) and rapidly became a new farming model VACB with B (biogas plant) as a key component for closing the energy and nutrient cycle in the farm. The first model is high cost, needs skilled labor, so that only about 3,000 units have been installed in the South of Vietnam. Meanwhile, the latter is much cheaper and easier to operate and flexible in its location so that up to now about 30,000 units have been installed in the South of Vietnam.

From 2002, VACVINA digester model has been designed and introduced by Center for Rural Communities Research & Development (CCRD). This model incorporating elements of concrete and PE technologies, and adapting them to the particular requirements and tastes of the Vietnamese market. Through the network of Vietnam's provincial Gardening Associations, around 10,000 VACVINA digesters were installed at famer's households who paid 100% investment cost (ETC/TaTEDO/CCRD, 2008).

From 2003, a cooperation project "Biogas Program for the Animal Husbandry Sector" was processed between Vietnam's Ministry of Agriculture and Rural Development (MARD) and the Netherlands Development Organization (SNV). With the commitment aid from Netherlands government, this project expects to construct 140,000 biogas plants up to the end of 2011. There is some of biogas plant models were introduced and constructed within this project such as KT1, KT2, KT3, etc. However, according to the hydrogeology condition of high groundwater table in the MD, only the KT2 biogas model introduced in the MD (Biogas Project Management Unit, 2013).

From 2006 to 2010, Luxembourg government support budget to do a project on "VIE020 - Sustainable production of aqua and agriculture and using renewable energy from water hyacinth and waste" in Hau Giang province. During the implementation of the project, scientists from Cantho University have designed, built, put into operation and successfully tested for two types of biogas plant namely EQ1 and EQ2. Both of these biogas plants were certified by Vietnam's MARD as an advance technique and were recommended "to quickly disseminate these achievements for contributing to diversify the biogas plant". By support from project, there are about hundred units of EQ1 and EQ2 are constructed in project area and other provinces.

In additional, in the MD also appeared some type of biogas models which developed by farmer themselves such as the water jar biogas plant, the septic tank biogas plant. Amount of these biogas models only few tens but it is shows the attraction of anaerobic technology to farmers.

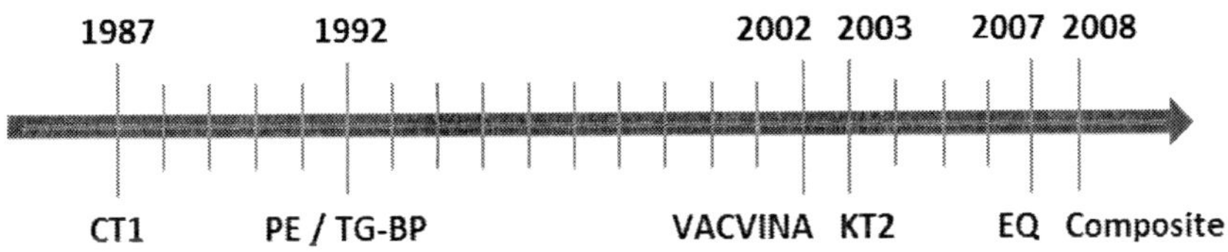

Figure 3. The appearing milestones of popular biogas plants in the MD.

1.2. Popular Biogas Plants in the MD

1.2.1. The PE Digester

This digester is made of a tube-like plastic digester bag in which the gas is stored in a separated PE bag. The inlet and the outlet are connected directly to the membrane of the tube. Since 1992, within the project supported by FAO and SIDA, the College of Agriculture and Forestry in Ho Chi Minh City introduced the PE tube biodigester (Bui et al., 1996). By low cost and simplicity of installation, this model has been rapidly adopted and disseminated by the network of the Agricultural and Forestry Extension - Vietnam Gardening Association (VACVINA) - and some local private activists (Le, 2008). In the MD, the PE digester is fabricated from transparent tubular polyethylene tubes of 0.16 mm thickness with a diameter from 0.8 to 1.4 m. The digester is made by overlapping two or three layers of PE tube together to increase the life time of digester. The length of a digester could be up to 12.0 m depending on the number of pig heads. To install the inlet and outlet pipes, two 0.15 m diameter pipes (ceramic or PVC) are inserted half of length to open ends of the PE tube, and fasten to PE. The joint is wrapped tightly with a rubber band similar to the inner tube of bicycles. A PVC pipe of 0.021 m diameter is fixed at one position on the upper surface of the digester bag and it is used as a gas vent. All the joints must be completely airtight. The digester is placed in a horizontal ditch with a slight slope from the inlet to outlet pipe. Three quarters of the digester's volume are filled with animal dung slurry through the inlet pipe and one-quarter is reserved for exhaust. Biogas is accumulated at the upper exhausting space of the digester and is flowing up through the gas vent into one separate gas-holder unit. The gas-holder bag, made of two layer of PE tubes so that the outside layer can protect the inside layer. The length of the gas-holder can be up to 6.0 m, according to the digester volume. Between the digester and the gas-holder bag is a safety water seal through which the biogas escapes if too much pressure is built up within the digester. To make this water seal, the farmer needs a 2-liter transparent plastic bottle (the used water bottle or the used cooking-oil bottle, etc.) filled with water.

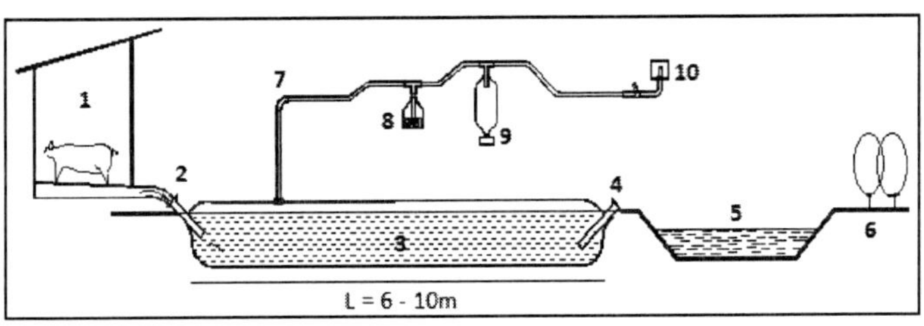

Figure 4. The PE digester. 1. Pig-pen; 2. Inlet pipe; 3. Digester; 4. Outlet pipe; 5. Discharge pond; 6. Garden; 7. Gas vent; 8. Security valve; 9. PE gas-holder; 10. Stove (Drawing: Vo H. N.).

A T-pipe was inserted into the bottle to the level of 0.08 ÷ 0.10 m under water. For this T-pipe, one side is connected to the inlet gas pipe coming from the digester, while another side is connected to the outlet gas pipe that leads to the biogas appliance. When biogas is produced in excess in the digester, a high-pressure will force the water in the safety water seal to empty the water inside the bottle. As a result, this action helps to release the extra biogas that maintains the stable pressure inside the digester and prevent the digester from being broken due to a high pressure.

As for the biogas filter, almost none of the farmers in the MD installed this unit into their biogas plants. The reason for this is the fact that the biogas is mainly used for cooking so that the farmers do not care of the negative effects of impurity gases (H_2S, CO_2) on their cooking tools.

The advantages of this model:

- Low cost, easy to install
- The fermentation slurry is agitated slightly by the movement of the digester body and the continuous-flow inside the digester
- Suitable for use in the MD where the groundwater table is high

The disadvantages of this plant:

- This channel-type digester takes more land area
- The PE layer is very sensitive to physical conditions such as the sunlight, falling objects, and by the interferences of people and animals so that they have short life-time than expected and can be easily damaged
- In case the sediment is accumulated in large amounts inside the digester, it is difficult to remove it
- By installation on the ground, the temperature inside the digester is unstable for decomposition process
- The produced biogas needs to be stored in a plastic bag separated from the digester body and it is not safe for this bag to be installed nearby a kitchen
- By storing the gas in a plastic bag, it has a rather low pressure (around 10 hPa) that only suitable to supply gas for cooking and not for lighting

Figure 5. The digester body (left) and the gas-storage (right) of the PE model.

1.2.2. The TG-BP Model

The TG-BP plant was developed within the Thailand-German Biogas Program from 1988 to 1995 in cooperation with the scientists from Germany (GTZ), and Thailand (Chiang Mai University and Department of Agricultural Extension). In 1992, this biogas plant was tested in Cantho province by REC, and then introduced broadly in Southern part of Vietnam. Up to now, there are about 3,000 units constructed in the MD, of which about 1,000 units installed in Sa Dec town - Dong Thap province and the rest built in other provinces such as Can Tho, Vinh Long, An Giang, Bac Lieu, etc. (Nguyen and Do, 2009).

There are three main components continuously connecting on this model: (i) the inlet tank where pig manure is mixed with water before it is discharged into digester; (ii) the digester where the mixture of pig manure and water is fermented to produce methane and other gases; and (iii) the displacement tank where the excess slurry is automatic overflow. These components act as a dynamic system. When gas is produced inside the digester, the gas pressure will push manure and slurry at the bottom of the digester to flow up into displacement tank. It could say the displacement tank is also creating the pressure for the gas in the digester body. When this gas is consumed, the slurry in the displacement tank will flow back into the digester to push the gas up for usage.

The digester made of solid bricks and mortar on surface of a concrete footing-beam having the hemispherical shape. A circular pit is dug into a depth base on farmer's yard. After the excavation is complete, the ground-soil has to be compacted. In the MD, due to the soft ground-soil and high groundwater table, piles are required at the foundation in case big biogas plants with volumes bigger than 18 m³. Due to its big weight, combined to big volume of liquid stored, the digester will be easily sinking if any crack appears in the foundation.

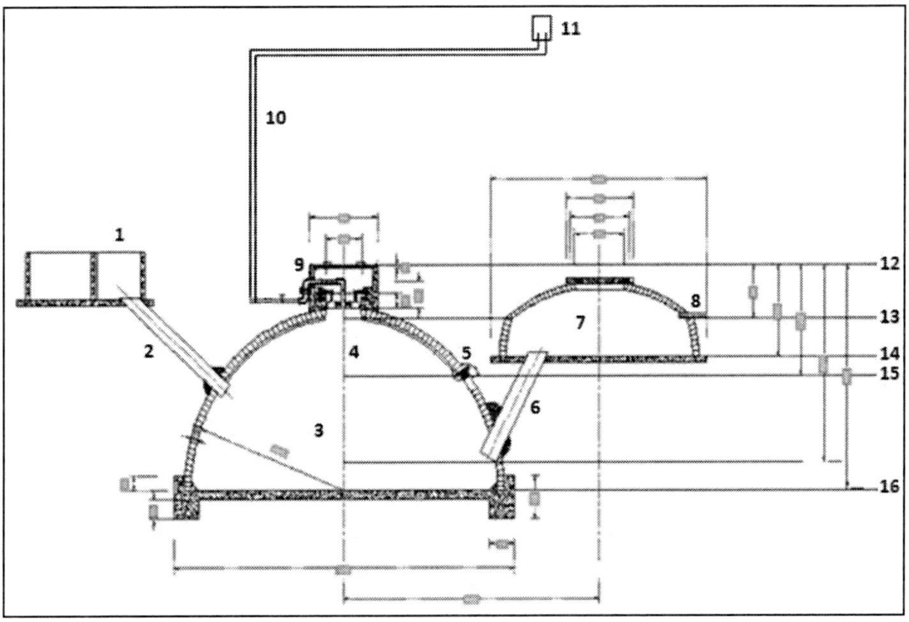

Figure 6. The construction drawing of TG-BP biogas plant. 1. Inlet tank; 2. Inlet; 3. Digester; 4. Fixed-dome gas-holder; 5. Soft ring; 6. Outlet; 7. Displacement tank; 8. Discharge pipe; 9. Entry hose with gas-light seal; 10. Gas pipe; 11. Stove; 12. Digester top level; 13. Ground level; 14. Zero line; 15. Soft ring level; 16. Digester bottom level (Source: Modify from REC, 2000).

The soil pit is cut out from both sides for inlet and outlet pipes. The mortar (one part cement and two parts sand) is spread over the surface of a stone layer (0.15 m thick) and compressed with a suitable quantity of water for making the foundations. A circular digester is constructed on the foundation by raising a brick wall of 0.1 m thick. For the brickwork, the radius stick must be used with each layer of brick to keep the digester radius constant. This procedure is slow since the workers have to wait the mortal getting dry in order to avoid cracks in the wall.

An inlet pipe 0.16 ÷ 0.21 m diameter made by PVC or cement is fixed in a slanting position with its lower end fitted in the digester wall above the foundation level. The upper end connects to the bottom of the inlet tank, which is normally connected to drainage canal of livestock pen.

An outlet pipe of 0.16 ÷ 0.21 m diameter is fixed to the opposite side with the inlet pipe. It is notice the outlet pipe diameter has to be larger than the inlet pipe diameter so that the excess slurry could easily to flow up to the displacement tank. In addition, the bottom edge level of the outlet pipe at the digester body should be lower than the bottom edge level of the inlet pipe that drains out the slurry at the bottom of the digester.

To avoid cracks on the digester body, two safety rings are attached on the digester wall: the soft ring and the hard ring. These rings are rather innovative when comparing to previous fixed-dome digesters. The soft ring is constructed to separate the bottom wall and the upper part of the dome.

In case the ground is sinking or the wall is cracked, this soft ring will prevent the vertical crack spreading up to the top of the dome. The soft ring is a flexible layer and is installed by placing bricks to form a wall outside the radius. The distance between the wall and the soft ring is one brick wide. The soft ring is 2 ÷ 5 layers mortal at the mixing ratio of 1 part cement, 3 parts lime and 15 parts fine sand.

The produced gas is stored inside the fixed-dome of the digester. This fixed-dome is located at the top of the digester and is separated from the lower wall by the soft ring. The fixed-dome is constructed from solid bricks by casting one layer of vertical bricks on top of the soft ring.

Figure 7. The construction of soft ring (left) and safety ring system (right). (Source: Modified from REC, 2000).

On each layer, the lower part of brick sticks out 0.03 ÷ 0.05 m towards the inner of the dome. The brickwork will continue until the top of fixed-dome is 0.38 ÷ 0.39 m wide. Three layers of mortar will plaster the inner of this fixed-dome. Each of the mortar layers is 1 cm thick and is made of mixed materials as following (i) the first layer by the mixture of 1 part cement: 1/3 parts lime: 2.5 parts fine sand, (ii) the second layer by mixing 1 part cement: 1/4 parts lime: 2.5 parts fine sand, and (iii) the third layer by mixing 1 part cement: 1/4 parts lime: 2.5 parts fine sand. These special mortar layers help to prevent the gas leakage through the fixed-dome. In additional, the mixture of 1 part cement: 1/3 parts lime: 2.5 parts coarse sand is required to plaster the outer wall until the thickness of 0.03 m is reached.

The hard ring is the beam of the fixed-dome and the protector of soft ring. The required strength of the hard ring is obtained by the mixing 1 part cement : 2 parts fine sand: 4 parts stones (stone size of 0.01 ÷ 0.02 m). This hard ring is formed of poured cement that covers the vertical bricks in a shape of mount.

The opening (0.38 ÷ 0.39 m diameter) at the top of the digester is used for maintenance works. Nevertheless, to ensure for air-tight conditions, a digester neck is built to support the lid. The digester neck is made of solid bricks and plastered mortar installed right on top of the digester opening to form a round-ring. A gas outlet valve (0.021 m diameter) is fixed on the lid of the dome.

The volume of gas-holder takes about 1/3 of the total digester volume. When the gas production commences, the slurry is displaced into the displacement tank through the outlet pipe. However, when the gas is consumed, the slurry flows back from the displacement tank into the digester, breaking the scum at the surface of slurry inside the digester. The displacement tank's dimension is according to the dimension of the digester with the volume of about 1/6 ÷ 1/8 of the digester volume. To build the displacement tank, the masons prepare a foundation at the zero line height level which must be at the same level as the lowest level of the slurry inside the digester. The foundation is constructed by a 0.05 m concrete layer obtained by mixing 1 part cement: 2 parts fine sand : 4 parts stones (stone size of 0.01 ÷ 0.02 m). The digester body is built completely underground while the digester neck is rising above ground level. Also the displacement tank is buried underground, too. This soil cover help to avoid any cracks and to let the weight of soil support against the uplift pressure of groundwater on the dome foundation. Vegetables can be plant on the top-soil to prevent eroding of soil.

Figure 8. The construction works on TG-BP model (Source: REC, 2000).

The advantages of this plant:

- The long life-time 20 years expected
- The mason already offer at least one-year warranty to this plant model
- The digester in form of semi spherical dome helps to eliminate construction failures while a digester constructed in rectangular form can lead to cracks at its corners
- The safety rings eliminates the failure of digester body
- Construction with available materials those are easy to find locally
- Saving land area due to the tanks are constructed almost completely underground.
- Less maintenance works and equipment replacement since the plant has less mobile equipment and no rusting steel parts
- High pressure of biogas (40 to 60 hPa) which enables the gas used as a substitute for solid/liquid fuels in cooking and lighting
- Due to underground construction, the slurry ferment is in stable conditions even in low temperature

The disadvantages of this model:

- The investment cost is high
- Requires skilled masons and needs supervision by experienced biogas technicians
- Unstable fermentation process due to changes on slurry level through the outlet pipe. When biogas is in use, the slurry containing oxygen from the displacement tank flows back into the digester causing adverse effect to anaerobic process. Due to this weakness, the TG-BP model mainly applies for small volume of digesters.

1.2.3. The VACVINA Model

The VACVINA digester model was developed in Vietnam to address specific issues faced by Vietnamese farmers. It is an innovative, low-maintenance and easily built model that is well adapted to the local context.

The underground flat-top system integrates both pigsty and latrine while providing a concrete floor on which the animal shelters are built. This reduces land requirements to a minimum (available space is a major hurdle for many households). On the other hand, its hybrid nature (plastic bag reservoir) helps keep the cost down. Furthermore, its innovative design prevents the accumulation of a hardened scum layer that reduces gas output in the absence of annual cleaning and overhaul.

The location of the digester normal chooses at under the cow/pigsty or animal sty due to digester body is flat concrete top. The concrete top can then become the pigsty or stable flooring, providing a dry and easy to clean surface. The shape of the digester does not necessarily have to be rectangular and can be designed to accommodate the disposition and topography of the family plot.

After selecting and covering the construction site, excavation is started. The dimensions of the pit must be larger than those of the digester so as to facilitate the execution of the work.

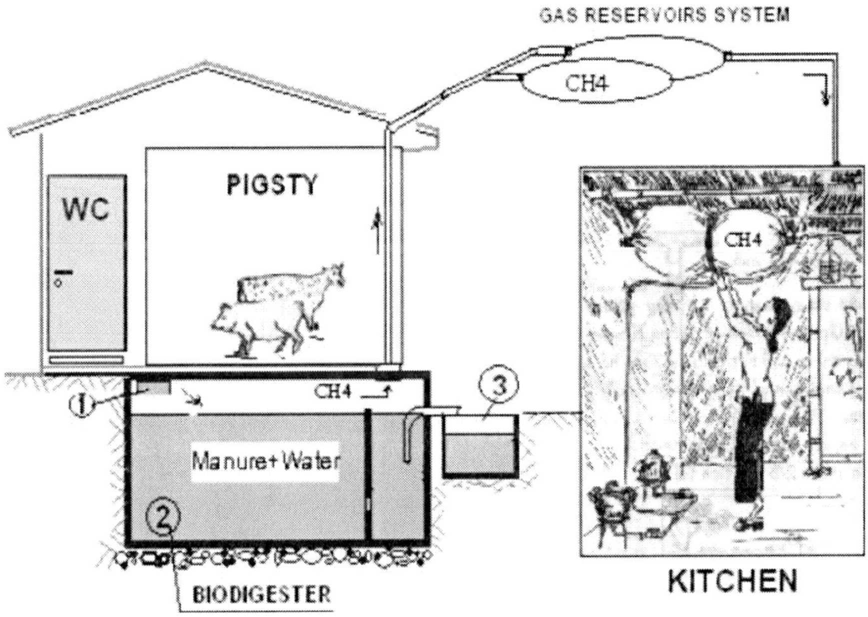

(Source: ETC/TaTEDO/CCRD, 2008)

Figure 9. The VACVINA biogas system.

The dimensions of the digester can be tailored to specific configurations. However, to minimize investment and maximize efficiency and ease of operation, it is suggested to keep the depth to 2.0 m, the width to 1.5 m ÷ 2.0 m and to adjust the length to obtain the total volume required. After completing the excavation of the pit, prepare a first 0.15 m bottom layer of tightly compacted broken bricks or crushed stone (0.04 × 0.06 m). Then pour a 5.0 cm concrete foundation using 1 part cement: 2 parts yellow sand: 3 parts crushed stone.

The walls are constructed after the base has been left to cure for at least 24 hours. The thickness of the walls is the standard 0.1 m (i.e. the thickness of a brick plus plaster). The bricks used must be solid grade A. The mortar is made of 1 part cement: 4 parts coarse sand. While laying bricks, ensure that the spaces (joints) between them are well filled with mortar. The plaster applied on the inner-side of the digester plays a very important role in ensuring permanent gas and water tightness. Plastering mortar is made of clean fine yellow sand mixed thoroughly with cement (1 part cement: 3 parts sand). Special care must be taken to ensure even thickness of the plastering layer. The surface must be smooth and regular and all edges must be rounded.

After plastering and polishing the walls, it is time to construct the digester cover. Because the flat-top cover is used to support the pigsty, animal stable and/or latrine, it must be at least 0.1 m thick, flat shaped and casted in place. The procedure is the same as that used to build a house foundation and comprises the following steps:

- Prepare a solid formwork for casting the cover. Use scaffoldings to support the wooden framework
- Define the location of the manhole to provide easy access to the digestion chamber while not interrupting animal production activities. Prepare a wooden mould shaped

as a truncated V-shaped rectangular column with upper dimensions of 0.5 m × 0.5 m, lower dimensions of 0.45 m × 0.45 m, and a height of 0.1 m

- Prepare a 0.15 m × 0.15 m mesh of ϕ8 steel rods, with hooks at both ends of the rods. Position the shorter steel rods under the longer ones for better solidity. If the width of the cover exceeds 2.5 m, a construction engineer must be consulted to design the proper reinforced mesh configuration
- Cast the concrete cover with 200-marked concrete: the basic 1 cement : 2 sand : 3 gravel mixture is used for casting the concrete cover with 200-marked concrete. To ensure even thickness of the cover, several 0.1 m wooden pegs are used. Once the thickness of the concrete layer is evenly compacted, remove the pegs.
- The main gas pipe must be fitted during casting of the cover. Made up of iron ϕ0.021 m, it is fitted right beside the wall of the animal stable so as not to disturb the animal production activities in the future
- Compact and sprinkle the concrete
- Casting a stove platform

This concrete platform will support the stove burner. It is made of 0.06 ÷ 0.08 m of concrete reinforced by a steel mesh of 0.2 m × 0.2 m ϕ8 rods with end-hooks. The breadth and length of platform are 0.6 m and 1.2 m respectively. For maximum convenience, it is recommended to install the platform 0.75 m above the kitchen floor. After completing the construction and installation of the system, all remaining excavated areas must be carefully filled back. The best material to use is sand, especially around the digester outer walls. The sand layers should be sprinkled with water and tightly compacted. The inlet system of the digester consists of a mixing tank connected to the main reservoir by one or more siphon pipes (also called U-band, rabbit-shaped joint, cattail-shaped joint). The mixing tank is usually located close to the digester and at the corner of the animal stable. The standard dimensions of the mixing tank are 0.6 m × 0.4 m with a depth of 0.4 m. These siphon inlets prevent the formation of a scum layer by regularly wetting and mixing the surface with fresh input. The number of inlets is based upon the size of digester, commonly 2 ÷ 3 per digester. The latrine can also be connected to the system. The outlet system consists of an outlet pipe and a slurry holding pit. The outlet pipe allows the digested slurry to flow from the digester to the slurry pit. Its length and position are important as they determine the level of slurry in the digester. The PVC pipe is 0.11 ÷ 0.15 m in diameter with a length equal to one third of the slurry height (0.8 ÷ 1.0 m). It must be positioned at an angle of 45°, making sure it extends under the surface of the slurry the equivalent of one third of the height of the slurry. The holding pit is commonly built alongside the outlet and its volume depends on the usage of the digested slurry. Install the gas pipeline connecting the gas outlet of the digester with the gas reservoirs and the burners. The safety valve is important to ensure that the gas pressure in the system does not exceed 0.15 m of water column. It consists of a 1.0 ÷ 1.5 liter plastic bottle, a T-joint and 0.021 m plastic tubing. To install the safety valve, make a hole in 0.015 to 0.02 m diameter at the neck of the bottle. Insert a short length of pipe into the bottle and connect the other end to the T-joint. Insert the T-joint as part of the main pipeline. Pour enough water into the bottle to cover the end of the pipe by 0.015 m. Mark the water level on the bottle as a guide to add water when necessary. If the pressure in the system reaches more than 0.015 m of water column, gas will automatically bubble out of the pipe and of the water bottle. Gas is stored in the gas storage until used for cooking. The plastic reservoirs are suspended under the stable roof or in an area where they are not subject to be punctured or affected by sunlight. The gas storage is made of two layers PE tubing, 1 m in diameter and 2.5 m in length. One end of bag is tightly tied while the other is connected to the pipeline with 0.021 m plastic tubing. The storage capacity of the reservoir is 1.8 m^3 and is enough for the 4 ÷ 5 hours of daily cooking required by a typical family of 5 ÷ 6 persons.

The advantages of this model:

- Less expensive than fixed dome systems
- Simple design (rectangular underground tank) and ease of construction, it does not require special masonry skills
- Possibility of pigsty or latrine directly on top
- Automatic scum breaking action
- Accommodates widely varying rates of gas production
- Very low maintenance

The disadvantages of this model:

- Above-ground piping
- Large storage reservoirs suspended in pigsty or kitchen
- It could be crack at the flat-top corners that cause of gas leakage out

1.2.4. The KT2 Model

The KT2 model is a further development version of the TG-BP model (Biogas project division, 2006). Together with KT1 and KT31, the KT2 model is developed within the project "Biogas Program for the Animal Husbandry Sector". Based on this national project, there are more than 50,000 KT2 plants constructed in Vietnam (Tran et al., 2009).

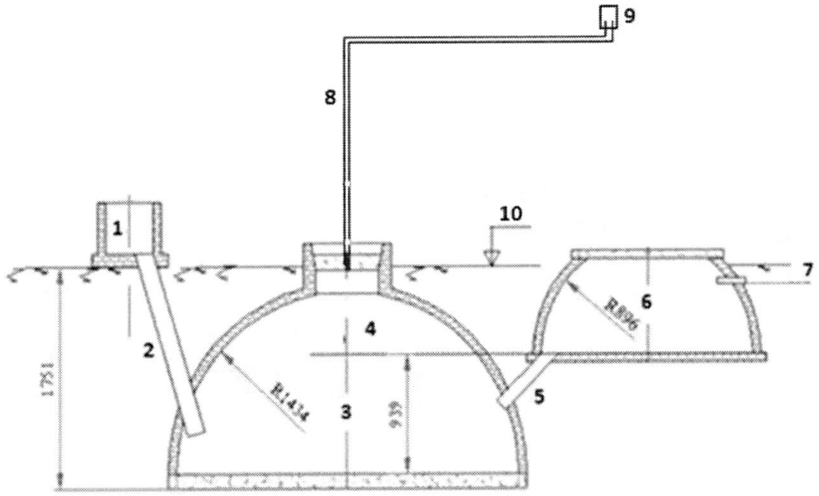

Figure 10. The KT2 domestic biogas plant. *1. Inlet tank; 2. Inlet; 3. Digester; 4. Fixed-dome gas-holder; 5. Outlet; 6. Displacement tank; 7. Over-flow level; 8. Gas pipe; 9. Stove; 10. Ground level* (Source: Modify from Biogas project division, 2008).

The characteristic of this model is the semi spherical dome on top of the plate bottom which makes the plant more resistant against the upward forces typically connected with high groundwater levels. Among KT models, the KT2 is, therefore, particularly suitable for the MD where the groundwater level is high. In the MD, about 7,254 units of KT2 biogas plant were installed at September 2010 (Le, 2010).

The volume of this digester is defined based on quantity of input material, dilute ratio of pig manure and water, needed volume of biogas, etc. Following the designed volumes of digester of 2, 3, 4, 5, 6, 8 and 10 m^3, the fixed livestock waste input to the digester will be of 50, 75, 100, 125, 150, 200, 250 and 300 kg per day.

The advantages of this biogas plant are, besides its strengths as mentioned in previous version (TG-BP model), the low investment cost comparing to TG-BP model due to its simple construction method on:

- No security rings system
- The digester is installed on a concrete floor without footing, which is required in TG-BP plant
- Is not using special mixing of materials like cement, sand and lime as required in TG-BP plant

The disadvantages of this model:

- The life-time of the KT2 model is presumed to be not as long as the life-time of the TG-BP model
- Without installation of the security rings package as the TG-BP model
- The mixing ratio water : pig manure is only from 1 : 1 to 1 : 2, which adverse effects to the treatment efficiency of the system

1.2.5. The EQ Models

The biogas plant models of EQ1 and EQ2 were developed by scientists from Cantho University within project "VIE020 - Water hyacinth" since 2008 (Nguyen and Do, 2009). These models represent an improvement and combination of TG-BP and PE models. Biogas produced from a concrete digester is stored in a PE bag for gas appliances. According to the project framework, the EQ1 and EQ2 models are mainly installed in Hau Giang province, where the project is implemented. After termination of the project, there have been about 75 units of both EQ1 and EQ2 built within the project in Hau Giang province and even more plants in other provinces of the MD. For the EQ1 plant, its digester body is similar to the TG-BP digester. Nevertheless, to save the cost of fixed-dome constructed by some special ways (plastering three mortar layers, preparing mortar with cement, sand, and lime), there is a HDPE circular cover layer fixed at the opening of the digester. Moreover, this plant is made of hollow-bricks (size of 0.08 × 0.08 × 0.18 m), saving the construction time and the labor cost. The biogas produced from fermentation process flows up to the gas-holder occupying about 1/5 ÷ 1/6 of the total digester volume. Then the biogas obtained moves out to a separate PE gas storage bag through a gas vent (0.021 m diameter) which directly connects into the HDPE cover layer. In addition, this EQ1 model includes no displacement tank so that the outlet slurry discharges directly into open-air place.

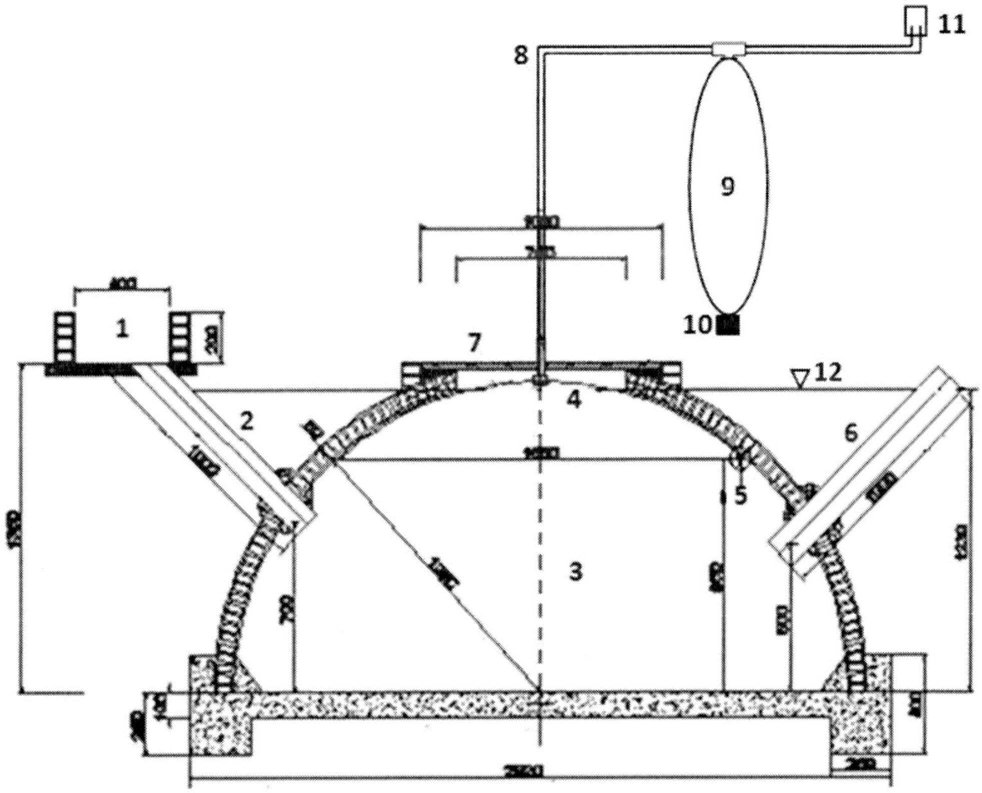

Figure 11. The EQ1 biogas plant. *1. Inlet pit; 2. Inlet; 3. Digester; 4. Gas-holder by HDPE; 5. HDPE hanger; 6. Outlet; 7. Cover plate; 8. Gas pipe; 9. PE gas-holder; 10. Weight; 11. Stove; 12. Ground level* (Source: Modify from Nguyen and Do, 2009).

Figure 12. The EQ1 biogas plant. *(left) the upper part of digester made by HDPE layer with gas collection point (right) the completed EQ1 plant with protect box and gas pipe* (Source: Nguyen Ngoc Em, Do Ngoc Quynh, 2009).

For the EQ2 model, the digester body seems to be similar to the digester body of EQ1 model. However, there is a closed fixed-dome installed on the top of the digester instead of the HDPE cover layer as installed in EQ1 model. At this closed fixed-dome, there is a crank handle hanging right at the top center and a gas pipe installed nearby. The crank handle is installed to diversity input materials. Farmers can apply some kind of biomass into the digester and then use this crank handle to breakdown the thick scum. This EQ2 model is also a replacement of the displacement tank with an open chamber with big hatch applicable for maintenance works.

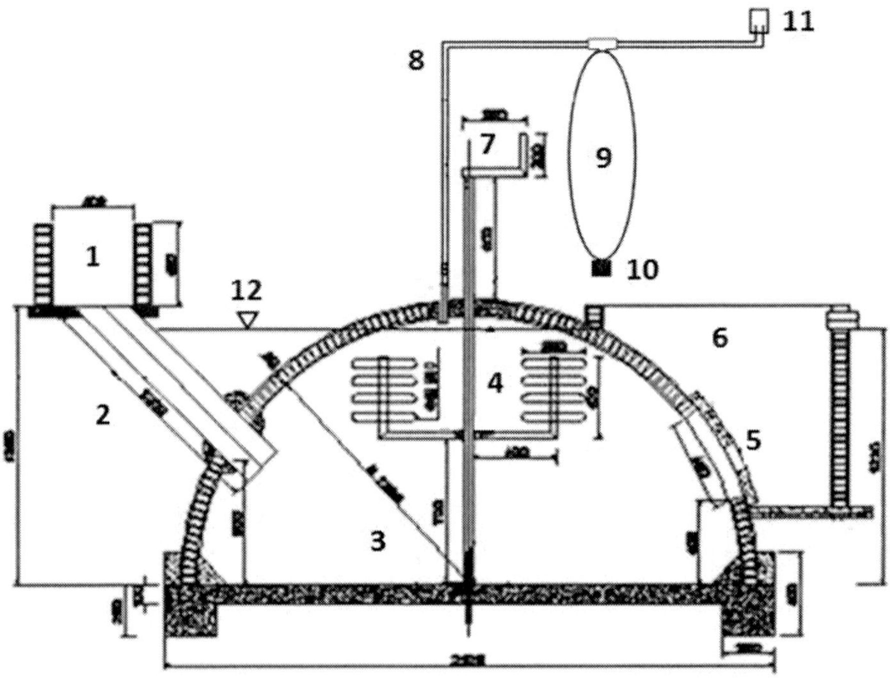

Figure 13. The EQ2 biogas plant. *1. Inlet pit. 2. Inlet. 3. Digester. 4. Stirring blades. 5. Discharge hatch cover. 6. Outlet pit. 7. Crank handle. 8. Gas pipe. 9. PE gas-holder. 10. Weight. 11. Stove. 12. Ground level* (Source: Modify from Nguyen and Do, 2009).

Figure 14. The EQ2 biogas plant. *(left) the completed EQ2 biogas plant with fulfilled outlet pit (right) the crank handle installed inside digester with two blades* (Source: Nguyen and Do, 2009).

The advantages of the EQ1 and EQ2 models:

- These models have separate gas storage bag so that the gas-holder volume inside the digester only takes about 1/5 ÷ 1/6 of the total volume of the digester while the gas-holder volume is 30 ÷ 40% of the total digester volume in TG-BP and KT2 models. As such, we can, in case of EQ models, save the labor cost and the materials cost.
- Saving construction time and labor cost due to the digester made of hollow bricks (size of 0.08 × 0.08 × 0.18 m compared to 0.04 × 0.08 × 0.18 m of the solid bricks).
- Due to the crank handle installed to EQ2 model, even biomass also input to digester. Furthermore it is convenient for maintenance work through a big hatch connecting to the digester and the open chamber.
- There is no requirement of skilled masons because these models do not require three mortar layers of cement, sand and lime, so that the local masons are often qualified enough to build these biogas models.

The disadvantages of the EQ1 and EQ2 models:

- A short life-time of the HDPE cover layer of the EQ1 plants. In addition, in case there is any block inside the EQ1 digester, farmers must drill a hole on the digester body for maintenance works.
- Regarding the separate gas storage PE bag in both EQ1 and EQ2 models, it is highly risky to keep this bag nearby the kitchen. Moreover, the gas pressure (from 0.05 ÷ 0.10 m of water column) is high only enough for cooking, not for lighting purpose.

1.2.6. The Composite Model

Composite plant is originated from China, being introduced to Vietnam since 2008. At present, this model has been manufactured by national companies. The biogas digester is made of synthetic material such as fiberglass, carbon fiber and polyester, which are 100% imported.

The composite model has two parts: the digestion part and the gas storage part, which is designed in one block and buried underground. In addition, the plant contains a fixed inlet for feeding material with the diameter of 0.8 m and a fixed outlet of 0.9 m diameter. Feeding materials are fed through inlet pipe until reaching the level of 0.6 m height of the inlet pipe (and also the outlet pipe).

At this time, the gas pressure in the digester is zero. After feeding, the inlet is close. The gas produced is stored in the upper part of the digester, and then it is pushed through gas pipe located in the middle point of the upper dome.

The digestion fluid is pushed through the outlet pipe. When the gas is used up, the gas pressure in gas section is 0. The biogas plant comes back to starting state. During the operation of fixed dome biogas plant, the substrate is moving up and down. The surface of the digestion solution narrows when is up and expands when it is down, which it will help reducing the formation of scum.

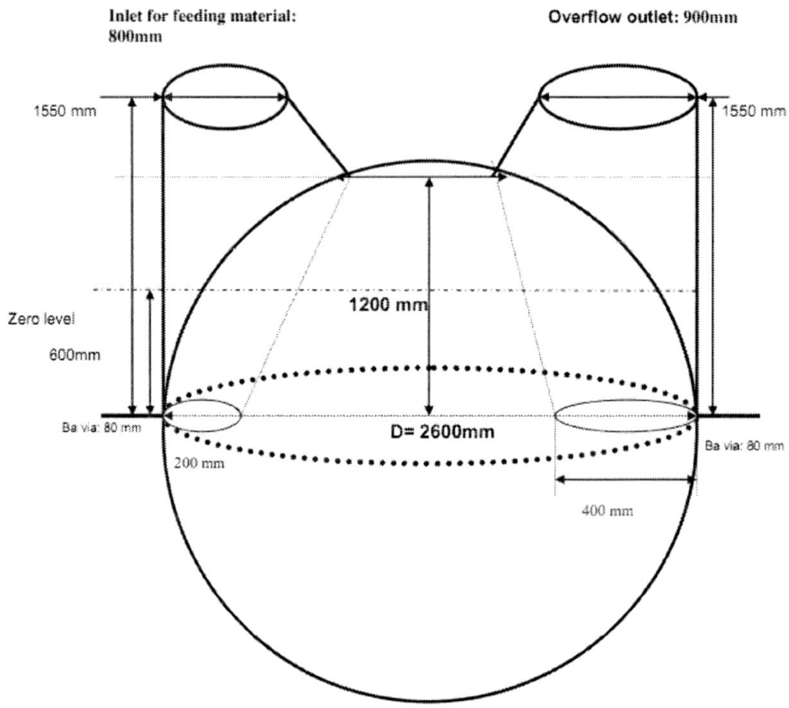

(Source: SEDDC, 2010)

Figure 15. The composite biogas plant.

Installation steps are summarized:

- Select site for digester
- Define main parameter (diameter of hole)
- Excavation: for selected sized of 9 cubic meters, the outer diameter of digester is 2.6 m. In the MD, normal the hole was digged with slopping of 1.0 : 1.5 due to high lever water table. The hole has parameter of 2.9 m × 2.6 m.
- Connect all parts of composite digester:

  - Connect ears: locate ear by marking on digester where to connect ears. Use prepare connection glue on brims of ear and digester then press glue on the surface of digester where connect ears. Mason goes inside digester and presses well all bolts and connect all bolts.
  - Drill a hole for gas pipe: use driller 10 to make a hole at the middle of upper digester wall. A drilled hole should have an inner diameter of 0.021 m. Press connection glue (like washer) and screw inner 0.021 m and screw outer 0.021 m together. Press well connection glue surrounding both inner screw and outer screw.
- Low composite digester into digged hole and fix by rod.

- Install inlet and outlet pipes: PVC pipes of 0.11 m were put on the ears of composite digester. Bricks were constructed to enclose the ears. Two concrete plates were put on the mouth of inlet and outlet to prevent rain water getting into digester.
- Install gas pipeline, gas meter and gas appliances; test water-tightness and gas-tightness.

Total time need for installation of composite model is 4 days in the MD. This time is more than the needed time to install one composite in the North (3.5 days) due to the special soil structure and the high groundwater table in this area. And due to high water table the composite digester floated after soil filling in the MD, thus the installation was repeated. To deal with this difficulty, a hole for collect stagnant water should be digged and a water pump is necessary to pump out water before installation of digester. Soil should be compressed. Sand bags may be used to put on. After that, water can be pumped into the digester to test water-tightness and gas-tightness.

The advantages of this model:

- This plant is gas-tight, water-tight and has high gas-pressure (up to 160 hPa), which makes possible the application of biogas for lighting.
- Saves construction area since the plant is buried under ground, suitable to weak soil.
- Saves time for installation, does not need trained masons because the composite installation was implemented by technician of suppliers.
- It can be moved to another location when necessary.
- Simple operation and maintenance works.

The disadvantages of this model:

- The big barrier implies high investment costs.
- It has big components that are difficult to transport to installation sites.
- Available in few digester volumes so that no much choices are available for the users.

This biogas plant was applied mostly in the northern and central part of Vietnam with more than 10,000 units (SEDDC, 2011). In the MD, this model just start spread since two past years, approximately hundreds units were installed.

Figure 16. Installing of the composite digester (left) and after soil filled (right) (Photographer: Vo H. N.).

1.2.7. The Improved Rectangular Parallelepiped Plant

This plant which developed and introduced by Soc Trang's Department of Agricultural and Rural Development is an upgraded version of VACVINA model (Le, 2005). This model is mainly installed in Soc Trang province with around 200 units.

The main improvement of this plant is that the inner side of the digester's flat-top is covered by a PE layer. This PE layer helps increasing the air-tight strength of the digester if any failure happens on the digester body.

This plant saves land area due to the digester being built right under a pigpen. The dimension of the digester is about 1.6 ÷ 2.0 m high, 2.0 m wide, and 3.5 m long. After the process of fermentation, the produced gas flows up to a separate gas storage bag. This bag is made of PE in 5.0 m length.

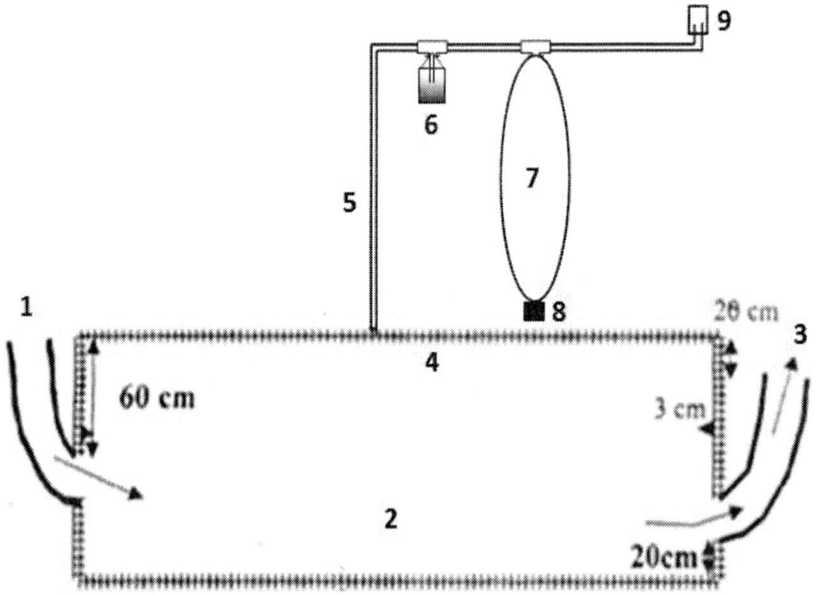

Figure 17. The improved rectangular parallelepiped biogas plant. 1 - Inlet pipe; 2 - Digester; 3 - Outlet pipe; 4 - PE layer; 5 - Gas pipe; 6 - Security valve; 7 - Gas storage. 8 - Weight; 9 – Stove (Source: modified from Le, 2005).

The advantages of this plant:

- The local masons are often qualified enough to build these biogas models.
- Easy to construct.
- Saves land area due to the digester being built under the pigpen.
- The leakage of the produced biogas through the cracks at digester corners is limited (better than VACVINA digester).

The disadvantages of this model:

- Low gas pressure in comparison to the semi sphere plants.
- Installation under the pigpen makes it difficult for maintenance works.

1.2.8. Farmer Self-Made Models

In the MD, besides the biogas plant models designed and applied within the biogas support projects, the local people, particularly where there is no beneficiary from the biogas projects and from the dissemination of biogas technology, have developed their own biogas plant models applicable for household scales according to their needs for gas supply and sanitation. Some of farmer self-made biogas models are:

- The biogas plant made from water jars.
- The biogas plant in rectangular parallelepiped shape.

a) The biogas plant made from water jars

This biogas plant model has been developed by the masons who made water jars. This model appears at the first time in 2009, up to now about 30 units have already been installed in Tra On district, Vinh Long province.

This biogas plant model is a system of two connected water jars in 1.3 ÷ 1.4 m height that are made from cement and fine stone. Each of the jars is 1.5 cubic meters volume, and the distance between the two jars is from 0.6 ÷ 0.7 m. The height installing level of the inlet pipe, the outlet pipe and a connecting pipe is 0.9 m from the jar bottom.

An outlet gas pipe is fitted on the jar cover plate. The produced gas flows to a separate gas storage PE bag.

The advantages of this plant:

- Low investment cost.
- Simple construction due to precast jars, quick construction time of 2 days.
- No transportation cost due to the pre-casting jars onsite.

The disadvantages of this model: by the height installing level of the outlet pipe and connecting pipe, much slurry accumulates in the jar, reducing the effective capacity of digester. In this case, it requires draining out the sediments from the digester frequently.

Figure 18. The water-jar biogas plant.

The total investment cost for this type of plant is approximately 2,585,440 VND. The unit investment cost of this biogas plant per cubic meter is $862,000 VN or $43.1 US (assumed exchange rate of 1 USD = 20,000 VND).

b) The biogas plant in rectangular parallelepiped shape

This biogas plant has been developed by the masons who have construction experiences. This model initiates in Long My - Hau Giang, but now it is spread to other areas. Local masons can build this plant based on the design of septic tank.

The dimension of the digester is about 1.6 ÷ 2.0 m height, 2.0 m width, and 3.5 m length. Its digester is divided into three chambers of which two big chambers and one small chamber. After the process of fermentation, the produced gas flows up to a separate gas storage bag. This bag is made by PE and 5.0 m long. There are two gas vents fitted on the top slab of the two big chambers.

This plant also saves land area due to the digester being built right under a pig-pen. The total investment cost of this plant is approximately $7,500,000 VN or $600,000 VN per cubic meter ($30.0 US/m³ with assumed exchange rate of 1 USD = 20,000 VND).

The advantages of this plant:

- Local normal masons are qualified to build this plant without any specific skills.
- Simple in construction works.
- Saving land area due to the digester being built underground.

The disadvantages of this model:

- Vulnerable to corner shattering due to the rectangular parallelepiped shape.
- Low gas pressure in comparison to the semi sphere plants.
- Large land area.

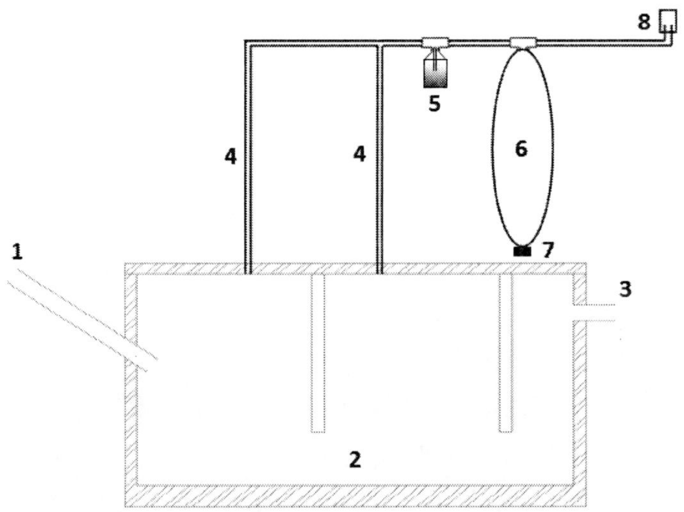

Figure 19. The sketch biogas plant in rectangular parallelepiped shape. 1. Inlet pipe; 2. Digester with three chambers; 3. Outlet pipe; 4. Gas pipes; 5. Security valve; 6. Gas storage; 7. Weight; 8. Stove.

1.3. Benefits of Biogas Technology

1.3.1. Methane Emission Reduction Potential

In general, biogas technology shows a cost-effective, environment friendly due to good waste management way. By digest the manure in anaerobic condition, biogas plant significantly reduce BOD, and pathogen levels; remove most noxious odors; and convert most of the organic nitrogen to plant available inorganic nitrogen.

In fact, the biogas plant itself well integrated to traditional VAC farming system in the Mekong Delta to establish a new eco-farming system VACB. In this eco-farming system, the output from each component will apply as input materials to other components, and vice versa.

Through anaerobic process, biogas was produced and applies mainly for cooking purpose, not for lighting or other energy purpose (Nguyen, 2012). In additions, bio-slurry form digester could apply as organic fertilizer to aqua-agriculture cultivation. There were some researchers implemented on benefits of biogas plant in the MD.

In the research on sustainable farming system model, Duong et al. (2000b) reported that the integration farming targeted to reduce investment and environmental management, hence it minimized risk from market price fluctuation. Studied on apply biogas sludge to tuber crops, Saleh et al. (2005) mentioned that the application of organic wastes can improve the performance of Al-tolerant while permitting the cultivation of more Al-sensitive crop on acid sulfate soils. Pham et al. (2010) reported that product yield of corn crop applied with biogas sludge better than applied with vermi-compost. Le and Duong (2010) defined water effluent from biogas digesters was used to improve aquaculture pond conditions, especially pH, in the low pH areas.

And in the interesting of promotion biogas technology in the MD, Nguyen (2012) did various experiments to find out benefits from biogas plant. He proved that water hyacinth and spent mushroom compost are potentially additional input for co-digestion with manure in biogas plants. In case of applying co-digester, the effluent helps not only increase the productivity of leaf mustard vegetable but also reduces the cultivation time thereof. Another benefit coming from applying the bio-slurry for gardening is an improvement in soil structure. In addition, the co-digester's effluent applied into the fishpond could increase the nutrient content essential for the development of algae and zooplankton which then feed for fish. The ideal benefit of biogas plant is reducing CH_4 emission to atmosphere.

In the MD, input material to biogas plant mainly comes from pig dung (Tran et al. 2009; Nguyen, 2012). According to GSO (2012), farmers in this area raised around 3.77 million pig heads in 2011. The average volatile solid (VS) production for per head per day of the average pig is 0.3 kg, and the maximum anaerobic bio-degradability of the VS is 0.29 m^3 CH_4 per kg VS (IPCC, 1997).

Base on these data, around 328,207 m^3 CH_4 generate per day in the MD if there is no treatment method for pig waste. There is no answer on how much of biogas plant was installed in the MD, but let estimate around 10,000 units were installed with average 6 cubic meter in volume of each. It means around 60,000 m^3 of biogas plant that could use to treat for 120,000 to 150,000 pig herds. This is a modest figure but by controlled capture of methane and its energetic use, part of methane can be eliminated in this area through existed biogas plants.

1.3.2. Environment and Sanitation Benefits

In the MD, there are 7 presented popular biogas plants including PE, TG-BP, VACVINA, KT2, EQ1, EQ2 and composite plants. As a matter of fact, although the quantity of installed EQ models are not much, just around hundred. But according to minute issued by MARD (2009), these models were approved as advance techniques so that they could involve to this list. Each of popular models has strengths and weaknesses in the construction, operation, and maintenance processes. The choice of a biogas plant model depends on the support projects, the local authority impacts, the economic condition of farmers, etc. An evaluation on these popular biogas plants in terms of their technical aspects and sanitation benefit as described below.

a) PE digester

Background:

- Input material: cattle manure
- Hydraulic retention time: 17 days
- Dilution ratio manure / water: 1 / 5
- The maximum loading: 6.86 kg/m^3*day^{-1}
- Gas pressure: 10 ÷ 20 hPa
- Investment cost: \$310,840 VN/m^3 ~ \$14.8 US/m^3 (refer to the construction unit price in Can Tho city in September 2010)

Table 1. Input and output concentrations from PE digester

Parameters	Unit	Concentration		QCVN 01 - 79:2011/ BNNPTNT
		Input	*Output*	
pH		7.0 ÷ 7.3	6.6 ÷ 7.0	6.0 ÷ 9.0
SS*	mg/L	2,700 ÷ 10,900	510 ÷ 2,800	50
Alkalinity	mg/L	2,880 ÷ 3,800	1,970 ÷ 2,980	-
BOD_5	mg O_2/L	2,180 ÷ 3,550	1,400 ÷ 2,200	30
COD	mg O_2/L	9,882 ÷ 32,000	4,608 ÷ 7,058	50
N total	mg/L	76 ÷ 3,970	42 ÷ 1,582	15
Total Coliform	MPN/ 100mL	$4.6*10^7$ ÷ $4.6*10^6$	$1.1*10^6$	3,000
Salmonella	MPN/ 100mL	$1.5*10^2$ ÷ $2.8*10^4$	$5.7*10^1$ ÷ $2.4*10^3$	-

Note: * refer to QCVN 40:2011/BTNMT (column A)

b) TG-BP plant

Background:

- Input material: cattle manure
- Hydraulic retention time: 20 days
- Dilution ratio manure / water: 1 / 5
- The maximum loading: 5.56 kg/m^3*day^{-1}
- Gas pressure: 40 ÷ 60 hPa
- Investment cost: \$1,277,800 VN/m^3 ~ \$60.85 US/m^3 (refer to the construction unit price in Can Tho city in September 2010)

Table 2. Input and output concentrations from TG-BP plant

Parameters	Unit	Concentration		QCVN 01 - 79:2011/ BNNPTNT
		Input	*Output*	
pH		6.8 ÷ 7.6	6.2 ÷ 6.6	6.0 ÷ 9.0
SS*	mg/L	6,750 ÷ 14,960	773 ÷ 1,000	50
Alkalinity	mg/L	1,015 ÷ 1,335	2,025 ÷ 2,325	-
BOD_5	mg O_2/L	1,516 ÷ 1,916	433 ÷ 917	30
COD	mg O_2/L	4,230 ÷ 7,239	1,411 ÷ 2,820	50
N total	mg/L	252 ÷ 555	133 ÷ 329	15
P total	mg/L	131.3 ÷ 285.7	113.7 ÷ 177.2	4
Total Coliform	MPN/ 100mL	$1.1*10^7 \div 4.2*10^7$	$2.3*10^3 \div 2.4*10^5$	3,000
Salmonella	MPN/ 100mL	$2.1*10^3 \div 1.1*10^4$	$1.2*10^3 \div 2.1*10^3$	-

Note: * refer to QCVN 40:2011/BTNMT (column A)

c) VACVINA plant
Background:

- Input material: cattle manure
- Hydraulic retention time: 45 ÷ 50 days
- Dilution ratio manure / water: 1 / 5
- Gas pressure: 10 ÷ 20 hPa
- Investment cost: \$450,000 VN/m^3 ~ \$22.5 US/m^3 (refer to the construction unit price in Can Tho city in September 2010)

Table 3. Input and output concentrations from VACVINA plant

Parameters	Unit	Concentration		QCVN 01 - 79:2011/ BNNPTNT
		Input	*Output*	
pH		6.5 ÷ 7.6	6.8 ÷ 7.4	6.0 ÷ 9.0
SS*	mg/L	6,200 ÷ 13,850	1,400 ÷ 4,450	50
Alkalinity	mg/L	1,300 ÷ 2,300	1,500 ÷ 4,100	-
BOD_5	mg O_2/L	900 ÷ 1,930	200 ÷ 270	30
COD	mg O_2/L	12,800 ÷ 42,352	4,510 ÷ 6,910	50
N total	mg/L	64 ÷ 2,394	50 ÷ 1,232	15
Total Coliform	MPN/ 100mL	$5.7*10^6 \div 5.7*10^6$	$1.1*10^5 \div 4.6*10^5$	3,000
Salmonella	MPN/ 100mL	$2.1*10^2 \div 2.3*10^3$	$4.6*10^2 \div 2.3*10^3$	-

Note: * refer to QCVN 40:2011/BTNMT (column A)

d) KT2 plant
Background:

- Input material: cattle manure
- Hydraulic retention time: 30 ÷ 40 days
- Dilution ratio manure : water: 1 : 2
- The maximum loading: 8.5 kg/m^3*day^{-1}
- Gas pressure: 60 ÷ 80 hPa
- Investment cost: \$1,279,113 VN/m^3 ~ \$60.91 US/m^3 (refer to the construction unit price in Can Tho city in September 2010)

Table 4. Input and output concentrations from KT2 plant

Parameters	Unit	Concentration		QCVN 01 - 79:2011/ BNNPTNT
		Input	*Output*	
pH		7.7 ÷ 8.7	6.6 ÷ 7.5	6.0 ÷ 9.0
SS*	mg/L	6,200 ÷ 8,825	440 ÷ 1,590	50
Alkalinity	mg/L	2,200 ÷ 4,000	1,700 ÷ 2,200	-
BOD_5	mg O_2/L	1170 ÷ 2930	980 ÷ 1,170	30
COD	mg O_2/L	9,600 ÷ 14,930	1,600 ÷ 2,130	50
N total	mg/L	229 ÷ 644	182 ÷ 294	15
Total Coliform	MPN/ 100mL	$2.4*10^6 \div 2.4*10^6$	$4.6*10^5 \div 1.1*10^6$	3,000
Salmonella	MPN/ 100mL	$2.0*10^3 \div 1.1*10^4$	$1.1*10^1 \div 2.0*10^3$	-

Note: * refer to QCVN 40:2011/BTNMT (column A)

e) EQ1 plant
Background:

- Input material: cattle manure
- Hydraulic retention time: 20 days
- Dilution ratio manure : water: 1 : 5
- The maximum loading: 6.67 kg/m^3*day^{-1}
- Gas pressure: 10 ÷ 20 cm hPa
- Investment cost: \$1,032,489 VN/m^3 ~ \$49.17 US/m^3 (refer to the construction unit price in Can Tho city in September 2010)

Table 5. Input and output concentrations from EQ1 plant

Parameters	Unit	Concentration		QCVN 01 - 79:2011/ BNNPTNT
		Input	*Output*	
pH		6.5 ÷ 7.7	6.2 ÷ 6.9	6.0 ÷ 9.0
SS*	mg/L	2,000 ÷ 2,000	1,033 ÷ 1,216	50
Alkalinity	mg/L	765 ÷ 2,000	1,858 ÷ 2,383	-
BOD_5	mg O_2/L	1383 ÷ 2100	330 ÷ 350	30
COD	mg O_2/L	5,030 ÷ 7,050	1,280 ÷ 3,330	50
N total	mg/L	233.3 ÷ 275.3	74.7 ÷ 81.0	15
P total	mg/L	235.9 ÷ 234.6	99.53 ÷ 147.97	4
Total Coliform	MPN/ 100mL	$2.4*10^8 \div 2.4*10^8$	$4.6*10^5 \div 2.4*10^6$	3,000
Salmonella	MPN/ 100mL	$3.0*10^1 \div 3.0*10^1$	$0.7*10^1 \div 0.9*10^1$	-

Note: * refer to QCVN 40:2011/BTNMT (column A)

f) EQ2 plant
Background:

- Input material: cattle manure + biomass (water hyacinth, rice straw, etc.)
- Hydraulic retention time: 20 days
- Dilution ratio manure : water: 1 : 5
- The maximum loading: 6.67 kg/m^3*day^{-1}
- Gas pressure: 10 ÷ 20 hPa

- Investment cost: $329,250 VN/m^3 ~ $63.30 US/m^3 (refer to construction unit price in Can Tho city in September 2010)

Table 6. Input and output concentrations from EQ2 plant

Parameters	Unit	Concentration		QCVN 01 - 79:2011/ BNNPTNT
		Input	*Output*	
pH		7.3 ÷ 8.2	6.8 ÷ 7.6	6.0 ÷ 9.0
SS*	mg/L	880 ÷ 5,800	54 ÷ 78	50
Alkalinity	mg/L	1,720 ÷ 2,680	2,950 ÷ 2,950	-
BOD_5	mg O_2/L	1150 ÷ 1300	270 ÷ 650	30
COD	mg O_2/L	8,320 ÷ 56,470	3,840 ÷ 9,880	50
N total	mg/L	148 ÷ 2,427	126 ÷ 1232	15
Total Coliform	MPN/ 100mL	$4.6*10^7 \div 4.6*10^8$	$2.4*10^6 \div 4.6*10^6$	3,000
Salmonella	MPN/ 100mL	$9.0*10^1 \div 9.3*10^4$	$1.4*10^1 \div 2.1*10^1$	-

Note: * refer to QCVN 40:2011/BTNMT (column A)

g) Composite plant
Background:

- Input material: cattle manure
- Hydraulic retention time: 10 days
- Dilution ratio manure : water: 1 : 1.5
- Gas pressure: up to 160 hPa (Moi truong xanh Composite, 2011)
- Investment cost: $2,613,333 VN/m^3 ~ $124.44 US/m^3 (refer to construction unit price in Can Tho city in September 2010)

Table 7. Input and output concentrations from composite plant

Parameters	Unit	Concentration*		QCVN 01 79:2011/ BNNPTNT -
		Input	*Output*	
BOD_5	mg O_2/L	33,290 ÷ 39,383	107 ÷ 195	30
COD	mg O_2/L	91,854 ÷ 99,450	234 ÷ 325	50
Total Coliform	MPN/ 100mL	$4.6*10^7 \div 4.6*10^7$	$2.8*10^4 \div 8.1*10^5$	3,000

Note:* source from SEDDC (2010)

Comparing some technical parameters and sanitation aspect of the popular biogas plants in the MD, there are some identifiers:

- The designed HRT of wastewater in the composite plant is shortest comparing to other plants. Inside the composite plant, the HRT of substrate is only 10 days compare to other plants from 17 to 20 days (except the KT2 model has HRT up to 30 to 40 days). The lower HRT, the lower of treatment efficiency.
- The designed dilute ratios manure : water differ from each model. While the discharge input to PE, TG-BP, VACVINA, EQ models apply the ratio 1 manure : 5 water, the KT2 and composite models apply 1 manure : 2 water and 1 manure : 1.5 water. But the composite model has the HRT only 10 days, there is no any guarantee for the effluent from this model reach the discharge standard.

- The models with separated gas-storage such as PE, EQ1 and EQ2 have low gas pressure and do not support farmers to apply the produced gas for lighting purpose. This is a weakness point of these models since the most of the rural areas in the MD lack electricity supply, especially in dry season.
- The EQ2 model with an innovation on crank handle is the sole model accepts multi-feeding materials. It is a strong point due to the insufficiency of pig manure that feeds into the biogas plant in the MD. Application of biogas plant with multi-feeding materials is a topic of DANIDA's funding project namely "Sustainable Production of Biogas from Waste Rice Straw" cooperation between Cantho University - Vietnam and Aarhus University - Denmark. The main objective of this project is to develop and test the feasibility of a simple and low-cost chemical/enzymatical process for converting waste rice straw to biogas at household scale (SubProM project, 2013).
- Investment cost is a big challenges to promote biogas plant in the MD (Nguyen, 2011). In this area, the PE digester has the lowest investment costs but also one of the shortest lifetime spans. In comparison, other concrete plants need higher investment costs (from 5 to 6 times higher), the composite plant has the highest investment cost. For household scale, at least a plant of 4 to 6 cubic meter need to install to treat for waste from around ten of pig, which popular raising scale in the MD. Therefore not much people paid themselves to build one biogas plant.
- Almost effluent from biogas plant contain high nutrient that not reach the national standard to discharge to open source. But by application biogas plant at the household scale, farmers have no responsibility to treat the biogas plant effluent in advance. In this case, the integrated farming system VACB shows great benefit on closing nutrient and eliminates pollution to surrounding water sources.

2. Anaerobic Digester for Industrial Wastewater Treatment

Over last decade, the Vietnam economy has rapidly developing with the industrialization and modernization process. The gross industrial export representing an annual average growth of 20 to 25%, in which aqua-agriculture industries playing an important role. This high economic development has increased living standard to rural communities where provide the aqua-agriculture raw materials for industrial processing.

However due to rapid economic growing, numerous serious environmental problems have been arise. Especially the untreated wastes from industrial zones result in tremendous environmental problems.

A common visible problem is the huge volume of wastewater discharge direct or indirect to received sources. In the MD, where produce more than 50% of national rice production and 60% of national aquaculture production, the high environmental pollution potential concerns the food enterprises, especially fish processing and brewery production.

2.1. Apply Anaerobic Digester to Treat Wastewater from Fish Processing Factory

2.1.1. Background

In the MD, due to annual export fish products about 70% of national fish product, fish processing is one of industry sectors which cause of serious pollution on environment. Fish processing activities generate potentially large quantities of organic components from inedible fish parts or endoskeleton shell parts from the crustacean peeling process.

The actual proportion depends on the edible fraction of each species being processed. Environmental effects from this industry sector are different due to production scale, final products, input materials, production technology, management level, etc. Among pollution sources of fish processing industry, wastewater is highly pollution source by its large volume discharge.

In which the industrial wastewater take in account from 85 to 90% total volume of issued wastewater. The main source of industrial wastewater from material washes, material processing, workshop and equipment cleaning. In fish processing factory, the different of raw materials and final products are caused of various water consumption (e. g. 5 ÷ 7 m^3 water per ton catfish, 4 ÷ 6 m^3 water per ton frozen shrimp, 20 ÷ 25 m^3 water per ton surimi, 4 ÷ 6 m^3 water per ton mixing frozen fish). The pollution level of wastewater depends on raw materials, final products, etc.

The wastewater from fish processing contain mainly bio-degradable organic components, nitrogen and phosphorus compound. As a result, the biological treatment technology is suitable to apply for this wastewater treatment. The popular systems for this treatment including anaerobic and aerobic combined processes such as UASB + activated sludge tank, UASB + activated sludge tank + A/O process.

Table 8. The concentrations of pollutant in fish processing wastewater

Parameters	Unit	Concentration		
		Frozen shrimp (1)	*Catfish (2)*	*Mixing (1) + (2)*
pH	-	6.5 ÷ 9.0	6.5 ÷ 7.0	5.5 ÷ 9.0
SS	mg/L	100 ÷ 300	500 ÷ 1,200	50 ÷ 194
BOD_5	mg O_2/L	500 ÷ 1,500	500 ÷ 1,500	391 ÷ 1.539
COD	mg O_2/L	800 ÷ 2,000	800 ÷ 2,500	694 ÷ 2,070
N total	mg/L	50 ÷ 200	100 ÷ 300	30 ÷ 100
P total	mg/L	10 ÷ 120	50 ÷ 100	3 ÷ 50
Fatty	mg/L	-	250 ÷ 830	2.4 ÷ 100

Source: VEA, 2011.

2.1.2. Case Study

To define the suitable wastewater treatment method which could apply to fish processing sector, VEA (2009) surveyed wastewater treatment systems at 11 fish processing factories. Each treatment system were ranked based on technology documents, onsite survey, and sample analysis results of three sampling times, there is 3 wastewater treatment systems were chose to promote for fish processing wastewater treatment. In which two systems applied anaerobic technology, detail of these anaerobic treatment systems shows below.

a) Factory 1

General information:

- Product: frozen fillet of catfish
- Factory capacity: 180 tons of raw material per day
- Wastewater volume: 3,600 m^3 per day
- Investment cost: 23,000,000,000 VND (in year 2008, $1 US ~ $20,000 VN)
- Operation cost: 3,600 VND per cubic meter

Table 9. The concentrations of pollutant in catfish frozen fillet wastewater

Parameters	Unit	Concentration		Treatment efficiency (%)	QCVN 11: 2008/ BTNMT (column A)
		Input	*Output*		
pH	-	7.0 ÷ 7.1	7.6 ÷ 7.7	-	6.0 ÷ 9.0
SS	mg/L	742 ÷ 795	4.0 ÷ 5.0	99	50*
BOD_5	mg O_2/L	1,803 ÷ 2,158	11 ÷ 27	99	30
COD	mg O_2/L	1,960 ÷ 2,320	20 ÷ 32	99	50
Ammonia	mg/L	22 ÷ 63	8 ÷ 13	41 ÷ 86	10
Residue Cl_2	mg/L	0.01 ÷ 0.02	0.05 ÷ 0.50	-	1
P total	mg/L	23 ÷ 41	4 ÷ 10	65 ÷ 87	4*
Total Coliform	MPN/ 100mL	6.0×10^5 ÷ 1.2×10^6	360 ÷ 2,900	100	3,000

(Source: VEA, 2011)

Note: * reference to QCVN 40:2011/BTNMT column A.

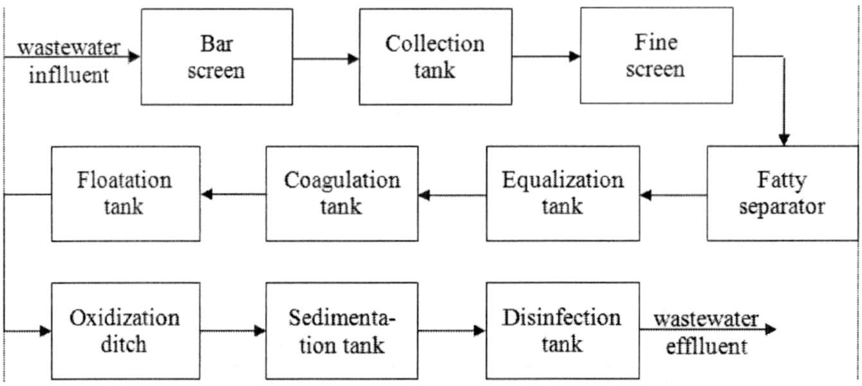

Figure 20. Wastewater treatment system diagram of catfish processing factory.

b) Factory 2

General information:

- Product: catfish fillet and catfish products
- Factory capacity: 70 tons of raw material per day
- Wastewater volume: 800 m^3 per day
- Investment cost: 6,000,000,000 VND (in year 2005, $1 US ~ $18,000 VN)
- Operation cost: 2,030 VND per cubic meter

Table 10. The concentrations of pollutant in catfish fillet and catfish products wastewater

Parameters	Unit	Concentration		Treatment efficiency (%)	QCVN 11: 2008/ BTNMT (column A)
		Input	*Output*		
pH	-	7.1 ÷ 7.1	7.6 ÷ 7.7	-	6.0 ÷ 9.0
SS	mg/L	456 ÷ 494	4.0 ÷ 8.0	98 ÷ 99	50*
BOD_5	mg O_2/L	1,491 ÷ 1,728	11 ÷ 30	98 ÷ 99	30
COD	mg O_2/L	1,600 ÷ 1,775	20 ÷ 32	98 ÷ 99	50
Ammonia	mg/L	24.7 ÷ 35.9	8 ÷ 25	28 ÷ 48	10
Residue Cl_2	mg/L	N/D ÷ 0.01	N/D ÷ 0.07	-	1
P total	mg/L	27.5 ÷ 52.3	3 ÷ 10	64 ÷ 94	4*
Total Coliform	MPN/ 100mL	$1.5\times10^6 \div 2.4\times10^6$	360 ÷ 2,600	100	3,000

(Source: VEA, 2011)
Note: * reference to QCVN 40:2011/BTNMT column A
N/D not detected

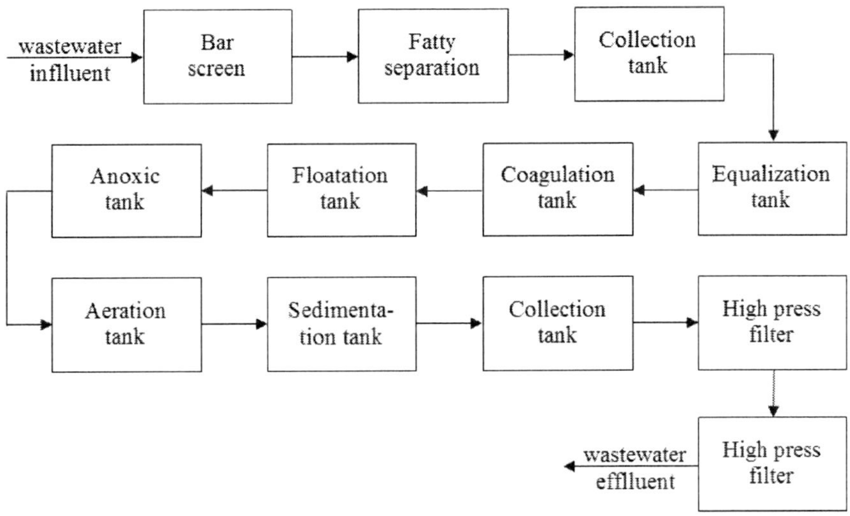

Figure 21. Wastewater treatment system diagram of catfish products factory.

2.2. Apply UASB Reactor to Treat Wastewater from Brewery Factory

In Vietnam, the beer market has been witnessing a high annual growth rate of between 9 to 11%. According to Japan's Sapporo beer producer, Vietnam is Asia's third-largest beer market after China and Japan (Vietnam Business, 2011).

The Vietnam government plan to develop the Beer - Alcohol - Beverage industry into a main economic sector (MoT, 2009). In fact, the market was increased 2.8 billion litters in 2011 regardless of economy situation.

From a beer factory, wastewater comes from three main sources including discharged water from washing the filter of beer wort preparation, the fermenters and bottles. Brewery effluent contains a lot of organic components such as sugars, soluble starch, ethanol, volatile fatty acids, etc. Besides, there are also biodegradable matter in the brewery solids including spent grains, waste yeast and hot trub.

Table 11. The concentrations of pollutant in brewery wastewater

Parameters	Unit	Input concentration	QCVN 40:2011/BTNMT (column A)
pH	-	6.5 ÷ 8.5	6.0 ÷ 9.0
DO	mg/L	0.5 ÷ 1.0	-
Color	Pt-Co	250 ÷ 350	-
Turbidity	FTU	120-150	-
SS	mg/L	250 ÷ 350	50
BOD_5	mg O_2/L	1,700 ÷ 2,700	30
COD	mg O_2/L	2,500 ÷ 4,000	75
$N\text{-}NH_3$	mg/L	12 ÷ 15	5
PO_4^{3-}	mg/L	20 ÷ 40	-

Source: Nguyen et al. 2008.

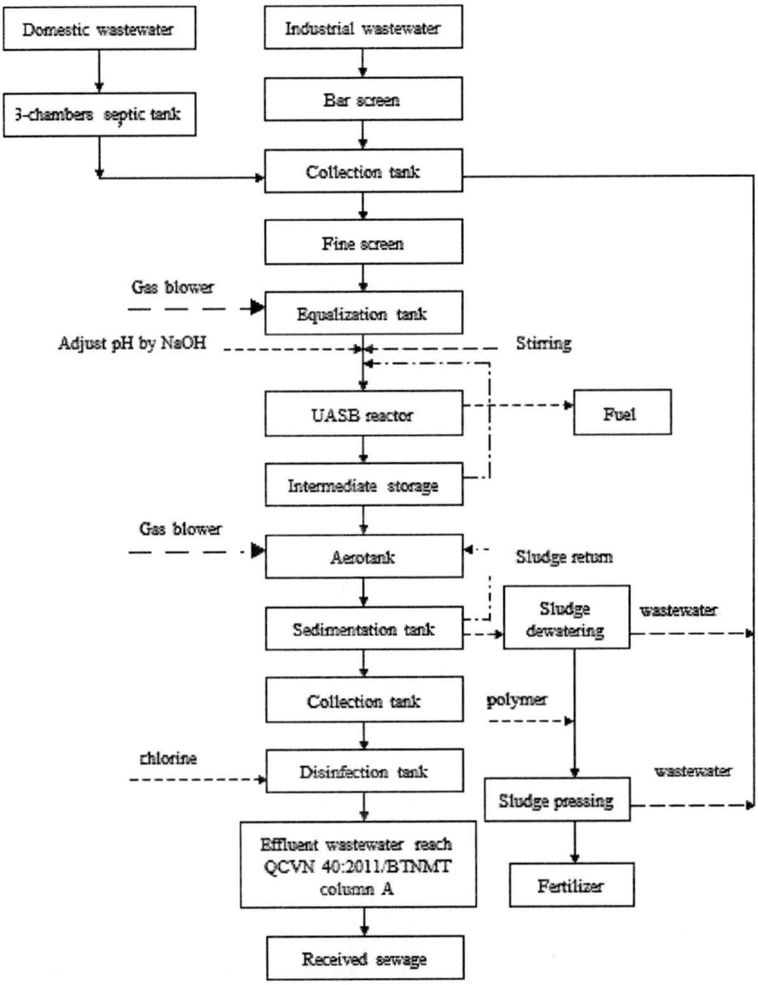

Figure 22. Wastewater treatment system diagram of brewery processing enterprises.

Because of high organic concentrations of soluble matter, the anaerobic treatment of brewery effluents prior to other treatments is an attractive method. In Vietnam the most popular anaerobic technology used in brewery enterprises is UASB reactor (Nguyen, 2007). Figure 22 shows a typical wastewater treatment diagram at one brewery factory in Can Tho City. In this system, there are two biological treatment processes which can be applied, anaerobic treatment (without oxygen) and aerobic treatment (with air/oxygen supply). In anaerobic treatment, organic matter will be converted into biogas while in aerobic process it will be oxidized into carbon dioxide and water. The produced methane from UASB reactor is not use for energy purpose, they just simple burn it. It is a big waste due to energy value of biogas in case using gas for heating or electricity generation.

CONCLUSION

Some of biogas plant models have been already constructed in the MD such as balloon model, TG-BP plant, KT2 plant, EQ1 and EQ2 plants, and some other self-made models. However, the dissemination of these models mostly depends on the biogas projects supported by some national or international agencies. The choice of biogas plant models applicable for the project areas is according to the project decision and each project normally chooses only one model of biogas plants to implement at the project area. This leads to a limitation in disseminating biogas application at large scale in these regions. It is because in case the farmers who are not beneficiaries from the projects but want to build a biogas plant cannot afford the biogas plant model that is selected by the project, these farmers have no alternatives for biogas plant models. It should be a good option for a biogas project to introduce some of other biogas models suitable and affordable for the local conditions to help the local farmers have more opportunities in approaching biogas technology. This introduction could offer some of low-cost models to attract more farmers to apply the biogas plant besides the actual biogas plant models applicable to the project area. By this way, biogas plants will be beneficiary to a larger number of local farmers as well as to the protection of the total local environment.

There is no "best choice" biogas plant model for all the farmers in the MD. The existing models available in the MD have their own strengths and weaknesses. For example, TG-BP model is long life-time and good operation but high investment cost while PE tube model is low investment cost but unstable operation and short life-time. It is required more studies to search for new optimal biogas models to the real situation of the MD of Vietnam. Such future biogas plant models should take three requirements into account including efficiency, economic, and large-scale application of the model when being designed for the MD. It is also worth of notice that most of the local people in the rural areas of the MD where biogas plants are in need are poor and low education. Cost saving is still a guiding principle in designing a biogas plant model applicable for the MD. Only by using cheap and available local materials, the low-cost biogas plant could apply at large scale in the MD. Moreover, a biogas plant should be simple in construction, operation, and maintenance. In respect of the diverse biogas usage and application, the gas pressure from the plant should be high enough for not only cooking but also lighting, and introduction of different models for different situations to give more alternatives for biogas application to the local people.

There is a limitation in information channels to introduce biogas technology at the rural areas. In case of no biogas plant installation, almost livestock waste discharges freely into open-air without any waste treatment methods. This causes more pollution on the open water source in the local areas that is used by most of the local people, especially the poor, for their living. Therefore, it is necessary to announce more communication programs on biogas technology, water and sanitation, etc. to rural residential communities in the process of improvement of the environment and livelihoods of the local people in the MD.

In order to push up the development of biogas technology in the MD, it is suggested that the benefit of reuse of bioslurry as a kind of fertilizer should be included beside of energy biogas and reduction of greenhouse gas emission.

REFERENCES

Biogas Project Management Unit, 2013. Project overview. Retrieved March 15th 2013, from http://biogas.org.vn/english/Gioi-thieu-du-an.aspx

Biogas project division, 2006. BP I final report. Biogas Programme for the Animal Husbandry Sector in some Provinces of Vietnam. Vietnam.

Bui Xuan An, Preston T. R. and Dolberg F., 1996. The introduction of low-cost plastic polyethylene tube biodigesters on small-scale farms in Vietnam. Retrieved September 15th 2010, from http://www.lrrd.org/lrrd9/2/an92.htm.

Do Kim Tuyen, 2005. Overview of biogas technology in Vietnam. Retrieved July 18th 2010 from www.methanetomarkets.org/expochina07/docs/postexpo/agvietnam.pdf.

Do Ngoc Quynh, 1989. Experience on organize and deploy the biogas plant at provinces in the Mekong Delta - The first step in Cantho city. In: *Biogas application in Vietnam*, edited by Lam Minh Triet et al. Proceeding on the first national workshop of the National Renewable Energy Programme 52C. Ho Chi Minh City, Vietnam. March 1989 (in Vietnamese).

Duong Van Ni, Ngo Thanh Binh, Huynh Van Nghiem, and Ryuichi Yamada, 2000. Farming system research on acid sulphate soils area of the Mekong Delta: Sustainable management. Retrieved September 15th 2012, from http://www.ctu.edu.vn/institutes/mdi/jircas/JIRCAS/research/workshop/pro03/F15-FS14(Dr.Ni).pdf.

ETC/TaTEDO/CCRD, 2008. Training and transfering VACVINA biodigester. Biogas Exchange visit program.

GSO - General Statistical Office, 2012. Statistical yearbook of Vietnam 2011. Vietnam's Statistical Publishing House (in Vietnamese).

IPCC - Intergovernmental Panel on Climate Change, 1997. Revised 1996 IPCC Guidelines for National Greenhouse Gas Inventories: Volume 1 Reporting Instructions. J.T. Houghton et al., IPCC/OECD/IEA, Paris, France.

Le Thi Huy, 2005. Method to build an improved biogas plant in cheap way. Retrieved July 19th 2010, from http://longdinh.com/default.asp?act=chi tiet&ID=707&catID=2.

Le Thi Phuong Mai and Duong Nhut Long, 2010. Use of biogas water effluent as material for improving aquaculture ponds. *Cantho University's Journal of Science*, No. 14b 43-55 (in Vietnamese).

Le Thi Thoa, 2010. Data on KT2 plant in the Mekong Delta. Biogas Project Division.

Le Thi Xuan Thu, 2008. Bio-slurry utilization in Vietnam. Report of Biogas Project Division. The Biogas Program for the Animal Husbandry Sector of Vietnam.

MARD - Ministry of Agriculture & Rural Development, 2009. Minutes of the Meeting of Scientific and Technical Council for Assessing the Technical Features of EQ Biogas Digester (in Vietnamese).

MARD - Ministry of Agriculture & Rural Development, 2011. Cattle and poultry farm - Hygiene inspection and evaluation procedure (in Vietnamese).

MoT - Ministry of Trade, 2009. Decision 2435/QD-BCT approved Development scheme for the Vietnam Beer - Alcohol - Beverage industry to 2015, vision to 2025 (in Vietnamese).

Moi truong xanh Composite, 2011. Strong point - weak point comparison of biogas concrete plant and composite plant. Retrieved July 19th 2012, from http://www.moitruongxanh composite.com/Tai-lieu-ky-thuat/bang-so-sanh-uu-nhuoc-diem-cua-ham-be-biogas-composite-va-ham-biogas-xay-bang-gach.html (in Vietnamese).

Nguyen Gia Luong, and Nguyen Quang Khai, 2005. Development of biogas technology in Vietnam. Retrieved July 19th 2010, from http://www.vcn.vnn.vn/PrintPreview.aspx ?ID=2895 (in Vietnamese).

Nguyen Ngoc Em, and Do Ngoc Quynh, 2009. Introduction on biogas plant models EQ1 and EQ2. Presentation at the final workshop of VIE/020-Water hyacinth project (in Vietnamese).

Nguyen Quang Khai, 1989. The fixed-dome biogas plant. In: *Biogas application in Vietnam*, edited by Lam Minh Triet et al. Proceeding on the first national workshop of the National Renewable Energy Programme 52C. Ho Chi Minh City, Vietnam. March 1989 (in Vietnamese).

Nguyen Van Phuoc, 2007. *Biological method for treating municipal and industrial wastewater*. Construction Publishing House (in Vietnameses).

Nguyen Van Viet, Tran Dinh Thanh, Pham Anh Tuan, Andrew Cikalov, and Thean Chong, 2008. Evaluation of the demonstration project at Hatay brewery in Vietnam. Retrieved May 19th 2013, from www.sdsap.org/data/hatay.pdf .

Nguyen Vo Chau Ngan, 2011. Small-scale anaerobic digesters in Vietnam - development and challenges. *J. Viet. Env*. 2011, Vol. 1, No. 1, pp. 12-18. ISSN: 2193-6471

Nguyen Vo Chau Ngan, 2012. Promotion of biogas plant in the Mekong Delta of Vietnam. PhD thesis, Technical University of Braunschweig, Germany.

Pham Thi Phuong Thuy, Ngo Thi Hong Tham, and Duong Minh Vien, 2010. Effects of different organic substrates on soil Al, Fe, P fractions and maize growth on acid sulfate soil. In: Clos*ing nutrient cycles in decentralized water treatment systems in the Mekong Delta. Can Tho city*, Viet Nam: Cantho University Publishing House (in Vietnamese).

REC - Renewable Energy Center, 1996. Demonstration presented at the Seminar for South East Asia on Anaerobic technology for waste and wastewater management and its economic, social and ecological impacts. Dec 1996, Ho Chi Minh city, Viet Nam.

REC - Renewable Energy Center, 2000. Biogas mason training manual for TG-BP plant construction (in Vietnamese).

Saleh E., Becker M., D.V. Ni, and K.V. Tinh. 2005. Biogas sludge reduces aluminium toxicity and improves tubers performance on acid sulphate soil of the Mekong Delta. *The Global Food & Product Chain - Dynamics, Innovations, Conflicts, Strategies*. Deutscher Tropentag.

SEDDC - Sustainable Energy Development Consultancy Joint Stock Company, 2010. Evaluation Study for Household Biogas Plant Models - Final Report. Retrieved May 19th 2012, from http://www.snvworld.org/en/publications/evaluation-study-for-household-biogas-plant-models.

SubProM project, 2013. *Introduction of SubProM project.* Retrieved May 19th 2013, from http://subprom.com.

Tran Viet Dung, Ha Viet Hung, and Huynh Thi Lien Hoa, 2009. Biogas user survey 2007 ÷ 2008. Biogas development programme for livestock sector in Vietnam.

VEA - Vietnam Environment Administration, 2011. Technical document - Guidelines for suitable evaluation of wastewater treatment technology and introduction of wastewater treatment methods to fish processing, textile industry, paper and wood-pulp production (in Vietnamese).

Vietnam Business, 2011. Vietnam beer market to increase 15% this year. Retrieved July 6th 2011, from http://vietnambusiness.asia/vietnam-beer-market-to-increase-15pct-this-year.

In: Anaerobic Digestion
Editor: Ana Torales
ISBN: 978-1-62808-884-7

Chapter 2

ANAEROBIC DIGESTION OF AGRICULTURAL WASTE: STATE OF THE ART AND FUTURE TRENDS

G. Merlin[*] and H. Boileau
Laboratoire Optimisation de la Conception et Ingénierie
de l'Environnement UMR
Université de Savoie, Campus scientifique Savoie Technolac
Le Bourget-du-Lac Cedex, France

ABSTRACT

Agricultural waste represents one of the largest anthropogenic sources of methane present in the atmosphere. Anaerobic digesters can turn these wastes into biogas as renewable energy, containing at least 50% of methane, and the solid residues into fertilizers rich in nutrients. Anaerobic digestion has a long history (10th century BC biogas used to heat water in Assyria) and many rustic digesters are implemented all around the world in farms. Unfortunately the lack of understanding and mistrust resulted in numerous failures, e.g. China and India have reported 50% failure rate and closure, 80% for US farm digesters. So, the knowledge upon anaerobic digestion limitations and reasons for failures is important; design and engineering deficiencies, complex digestion processes and inappropriate equipment installed… There is no standard designed digester for different reasons: in one hand, the situations are very diverse in operating conditions and in the other one, there remain many research issues still incomplete concerning microbiology, microbial ecology, metabolic pathways, modelling, technology… First, this chapter will be a recall of basic principles of anaerobic digestion applied to this kind of complex substrate. A review of the main types of agricultural waste digesters used around the world, from the simplest to the most sophisticated will be made by presenting advantages and disadvantages of each including environmental impact. It will also compile state-of-the-art information related to new developments in anaerobic technology applied to agricultural waste treatment.

But not only high technology could help in a correct control of the installation, and thus, a higher biogas production rate, also a balanced ration is required, and some simple recipes could help. Another aspect will be studied, namely the wishes of the manager of

[*] www.locie.univ-savoie.fr.

the facility. Indeed, depending on whether one wishes above all to produce more biogas (i.e. energy, therefore financial income), essentially to treat its effluents or produce a digestate agronomically interesting, driving of the installation and operation applied may differ.

INTRODUCTION

Agricultural waste is waste produced at agricultural premises as a result of an agricultural activity. An agricultural establishment produces many types of waste during its daily operations. Those are liquid or solid waste that result from agricultural practices, such as cattle manure, crop residue (for example, corn stalks), pesticides, and fertilizers (McGraw-Hill Dictionary of Scientific & Technical Terms, 2003).

Agricultural waste has become an increasing concern in recent years, as they may cause significant environmental problems; however, they may also be used for several beneficial purposes, as feedstock for energy production (Zhang *et al.*, 2012). This is particularly true in some countries, e.g. India that is bestowed with vast livestock wealth and it is growing rate of 6% per annum. Efficient utilization of by-products has a direct impact on the economy and environmental pollution of the country. Non-utilization or under utilization of by-products not only lead to loss of potential revenues but also lead to the added and increasing cost of disposal of these products.

Pig manure constitutes the largest part of animal agro waste in most industrialized countries, and cow manure is the main component in many developing countries especially in South-East Asia (Wint and Robinson, 2007).

Anaerobic digesters can turn these wastes into biogas as renewable energy, containing at least 50% of methane, and the final solid residue into fertilizers rich in nutrients. Anaerobic digestion is a promising valorization technology due to its ability to convert almost all sources of biomass, including different types of organic waste, slurry and manure to a highly energetic biogas (Holm-Nielsen *et al.*, 2009).

Anaerobic digestion has a long history (10th century BC biogas used to heat water in Assyria) and many rustic digesters are implemented all over the world in farms. 30 million units of biogas tanks have been installed in China with 6 million units between 2000 and 2005 (as of 2005) (Lin *et al.*, 2011).

In the UK an estimated 90 million tons of manures and slurries are generated each year (http://www.environment-agency.gov.uk/business/sectors/32601.aspx).

De Meester *et al.* (2012) study the environmental sustainability of anaerobic digestion from three perspectives. First, reference electricity is compared to electricity production from domestic organic waste and energy crop digestion. Second, different digesters feed possibilities in an agricultural context are studied. Third, the consequences of applying digestate as fertilizer are investigated. Results highlight that biomass is converted at a rational exergy (energy) efficiency ranging from 15.3% (22.6) to 33.3% (36.0). From a life cycle perspective, a saving of over 90% of the resource is achieved in most categories when comparing bio-based electricity to conventional electricity. Anaerobic digestion performs well in most categories by inducing significant resource savings. The digestion of agricultural residues such as manure and corn stover is favourable. To increase environmental sustainability, it is necessary to control emissions in the biogas production chain. Overall, unlike many other conversion technologies, anaerobic digestion is able to produce biogas from many different sources of biomass whilst co-producing a rich digestate.

Extensive production of agricultural crops and livestock in the rural areas simultaneously provides abundant amounts of residues every year and particularly in Asia (Gadde et al., 2009; Hossein and Badr, 2007; Ravindranath et al., 2005). Although often utilized, much of the residues are left unused. They have diverse characteristics and they stay in scattered places. Rahman and Paatero (2012) propose a method to quantify the residues with so-called residue characteristic factors data along with their processing techniques. This method determines the energy potential of the residues for anaerobic digestion processes instead of combustion, and correlates data directly with annual crop yields and livestock productions. The proposed method has been applied to five South Asian countries for estimating their potentials. The results show that the available agricultural residues in the rural areas of Bangladesh, India, Nepal, Pakistan, and Sri Lanka have annual electricity production potentials of 680 kWh, 820 kWh, 720 kWh, 1200 kWh and 480 kWh per rural household (HH) respectively. In such countries, village-scale utilization is found to be financially attractive (having per unit electricity generation cost 0.040 €. kWh^{-1}) to utilize the residues for electricity generation.

Large quantity of crops residues are produced annually worldwide and often dumped in open environment. These include cereal crops straw, rice husk, bagasse, maize cobs, coconut husk, groundnut and other nutshells and sawdust (Raphs, 2004). The above estimate of annual biomass production from maize, wheat, rice and sugarcane crops yields is about 5358.5 million tons of dry biomass. Though, its major part is utilized as feed for animals, but a huge amount of this biomass remains unutilized and or is burnt in open field environment. Rice husk and sugarcane bagasse are usually accumulated in large volumes at one site. Rice husk is among the commonest agricultural residue. It is well known that large cities surrounded by large agricultural areas use to suffer from rice husk burning in open air, as e.g. in Cairo. Rice husk amounts 20–25% of the harvested rice grain on dry weight basis, which is usually separated from rice grain at the processing centres (Chandra et al., 2012).

Agricultural waste does not compete with food crops in agricultural land usage.

Generalities on Agricultural Waste and Anaerobic Digestion

Anaerobic digestion (and/or co-digestion) is an effective and environmentally attractive pathway and promising option for recycling agricultural by-products because these contain high levels of biodegradable materials.

Compared to aerobic treatment, anaerobic process saves the energy needed for aeration, generates much less sludge (almost 50 times less), and also produces an energy-rich biogas (containing at least 50% methane). Anaerobic degradation of organic matter may be divided into three steps conducted by *Bacteria* and *Archaea* microbial species. In the first step, hydrolytic bacteria degrade polymeric organic matter into monomers (i.e. sugars, amino acids). During the second step, monomers are degraded into volatile fatty acids (e.g. acetate, formate…). In the third step, acetic acid, formate and hydrogen are reduced into carbon dioxide and methane by acetotrophs, methylotrophs and hydrogenotrophs bacteria. The anaerobic digestion can also be divided in four phases for the biomass degradation and conversion, namely hydrolysis, acidogenesis, acetogenesis, and methanogenesis (Figure 1). The partial pressure of hydrogen in acidogenic and acetogenic phases regulates acetate formation.

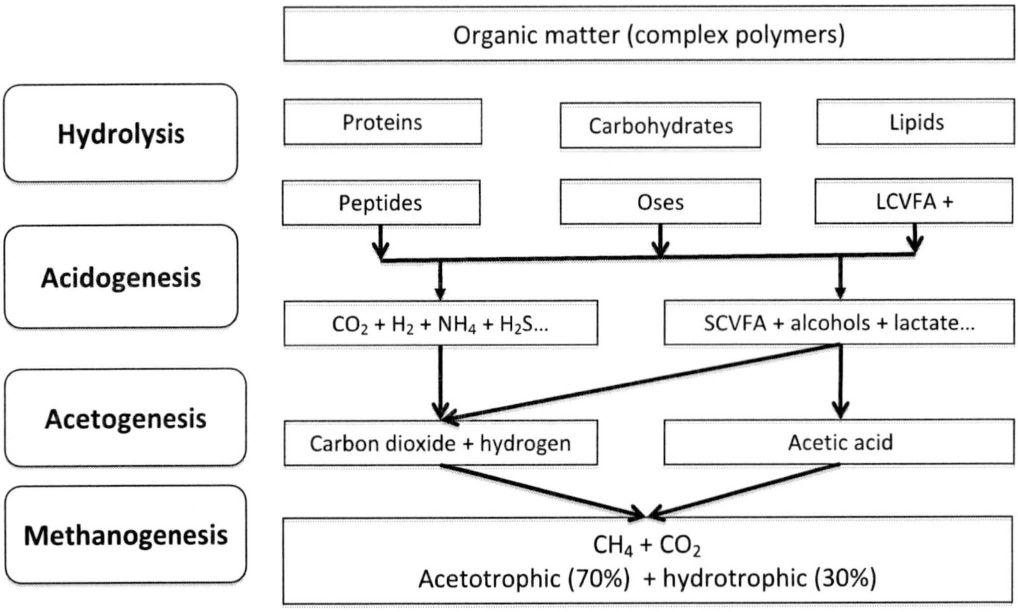

Figure 1. Schematic description of anaerobic digestion process. Long-Chain Volatile Fatty Acids (LCVFA), Short-Chain Volatile Fatty Acids (SCVFA).

Biogas production depends on different factors including: pH, temperature, substrate, loading rate, hydraulic retention time (HRT), C/N ratio, mixing…Main environmental optimal conditions required for anaerobic digestion process are summarized in table 1. Digestion is a slow process and it takes a minimum of 15-20 days for the microorganisms to adapt to new growing conditions when occurs a change in substrate composition or temperature level. (Deublein and Steinhauser, 2008).

Among the various factors affecting production of biogas, the temperature is the most important as the methanogens bacteria are very sensitive to sudden thermal changes. Investigation on the sensitivity to temperature variations has been achieved on a full-scale experimental dairy wastewater treatment plant including an unheated but insulated upflow anaerobic sludge blanket (UASB) (Merlin et al., 2012). In this study, large quantities of NaOH have been used to maintain the pH value of the influent to the digester at a value of 6.8. Average daily consumption was 42 kg of NaOH (0.105 $m^3 .d^{-1}$ of 30.5% NaOH water solution). This abnormal NaOH consumption is in relation with the low temperature in the digester that was close to 20-23°C. BMP tests have confirmed the influence of temperature on the bacterial activity (biogas production rate) decreasing by values ranging from 1.63 to 1.98 each 10 °C drop below 37 °C.

The pH of the system depends on the rate at which intermediates are formed during the fermentation process and the anaerobic digestion process is also very sensitive to this pH evolution. In a properly operating anaerobic digester, a pH value between 6.8 and 7.2 is observed as volatile fatty acids are converted to methane and carbon dioxide.

Table 1. Optimal conditions for biogas production

Parameter	Hydrolysis/Acidogenesis phase	Methanogenesis phase
pH	5.2 - 6.3	6.7 – 7.5
C/N ratio	10:1 – 45:1	20/1 – 30/1
Temperature (°C)	25 – 35	32 - 42 (mesophilic) 50 – 58 (thermophilic)
Redox potential (mV)	+400 to -300	< -250

Adapted from Chandra *et al.*, 2012.

Anaerobic digestion is a microbial conversion method that occurs in an aqueous environment, meaning that biomass sources containing high water levels can be processed without any pre-treatment (Ward et al., 2008).

Anaerobic digestion is a complex and multi-step microbial process, in which the cooperation between key members of the microbial community is necessary to optimize performance of the digestion (Buffiere et al., 2008).

Biogas yield and composition of the biogas are greatly affected by the composition of feed materials in respect of carbohydrate, fat and protein contents. This is verified by looking at the theoretical biochemical methane potential calculated from the elemental chemical formulas (equations 1 to 4).

Carbohydrates:

$(CH_2O)_n \rightarrow 0.5\ n\ CH_4 + 0.5\ n\ CO_2,\ Y_{CH4} = 0.37\ Nm^3\ .kg^{-1}$ sugar (eq. 1)

Proteins:

$C_{3.12}\ H_{4.94}\ ON_{0.82}\ S_{0.03} + 2\ H_2O \rightarrow$

$1.61\ CH_4 + 1.51\ CO_2 + 0.82\ NH_3 + 0.03\ H_2S\ ;\ Y_{CH4} = 0.51\ Nm^3\ .kg^{-1}$ protein (eq. 2)

Lipids:

$5.62\ CH_4 + 2.38\ CO_2\ ;\ Y_{CH4} = 1\ Nm^3\ .kg^{-1}$ fat (eq.3)

Biomass:

$C_5H_9O_{2.5}NS_{0.025} + 2.26\ H_2O \rightarrow$

$2.62\ CH_4 + 2.38\ CO_2 + NH_3 + 0.025\ H_2S\ ;\ Y_{CH4} = 0.48\ Nm^3\ .kg^{-1}$ VS (eq.4)

It is possible to use general formulas (equations 5 and 6):

$C_1H_aO_bN_c+(1\text{-}0.25a\text{-}0.5b+0.75c)H_20 \rightarrow$

$(0.5+0.125a\text{-}0.25b+0.75c)CO_2+(0.5+0.125a\text{-}0.25b+0.75c)CH_4 + cNH_3$ (eq. 5)

$C_cH_hO_oN_nS_s + 1/4\ (4c - h -2o + 3n + 2s)H_2O \rightarrow$

$1/8(4c - h -2o - 3n - 2s)CO_2 + 1/8(4c - h - 2o - 3n - 2s)CH_4 + nNH3 + sH_2S$

(eq. 6, Symons & Buswell, 1933)

When digesting substrate or wastes with unknown organic compounds, the ratio between the chemical oxygen demand (COD) and total organic carbon (TOC) allows predicting the CH_4 content according to:

CH_4 in % = 18.75 COD/TOC (eq. 7, Van Lier et al., 2008)

These complex microbiological pathways are a major advantage in comparison to other forms of bio-fuel production due to better conversion efficiency of biomass to biogas (Börjesson and Tufvesson, 2011).

The water content in substrate should be about 90% of the total weight. The water content differs according to the raw material used for fermentation. In case of fresh cattle dung, it has to be mixed with water on a unit volume basis (1/1, water /dung) before feeding into the digester (Chandra et al., 2012).

The organic loading rate (OLR) is defined as the amount of volatile solids (VS) or chemical oxygen demand (COD) components fed per day per unit of digester's volume. A total solid concentration of 15.2% has been reported to lead to the highest yield of gas production in case of dairy manure. Furthermore, it has also been found that in case of fresh dairy manure, a feedstock having total solids between 13.0% and 15.0% appears most ideal (Chandra et al., 2012).

Maintaining a suitable carbon to nitrogen ratio is essential for sustainable digestion. A C/N value of at least 25/1 is suggested for optimal gas production (Gerardi, 2003; Hartmann and Ahring, 2006). Animal waste, particularly cattle dung has an average C/N ratio of about 24.0. Vegetal materials such as cereal straw and sawdust contain a higher percentage of carbon. Materials with high C/N ratio could be mixed with low C/N ratio materials.

Crops Waste Digestion

Inedible residues from food crops (e.g. leaves and vegetable waste) and dedicated energy crops residues (e.g. maize, beets and wheat) offer a large potential for anaerobic digestion.

Plant biomass is mainly composed of lignocellulosic compounds (cellulose, hemicellulose, and lignin) which are the main constituents of cell walls. Cellulose is a linear polysaccharide polymer of glucose (~□1.4-glucan). Hemicellulose consists of shorter chains – 500~3,000 sugar units as opposed to 7,000~15,000 glucose molecules per polymer seen in cellulose. Hemicellulose is a branched polymer while cellulose is unbranched. Lignin is a very complex large molecule made of phenolic monomer units linked in a three-dimensional structure. In general, herbaceous plants such as grasses have the lowest contents of lignin, whereas soft- woods have the highest content of lignin (Chandra et al., 2012). Lignin makes lignocellolosic materials resistant to chemical and biological degradation (Perez et al., 2002). The rate limiting step for anaerobic digestion of lignocellulosic material is the hydrolysis of cellulose and hemicelluloses into mono-, di- and oligosaccharides. Increasing the hydrolysis rate is critical in order to improve the biomass-conversion efficiency in anaerobic digestion (Bruni et al., 2010).

Cellulose containing material like straw from wheat, rice and sorghum are widely available as a waste product from food production (Table 2). The potential methane yield of cellulose-containing materials in most cases can only be achieved after pre-treatment because of the high proportion of non-degradable materials present (Petersson et al., 2007; Weiland, 2010). Diverse pre-treatment techniques such as thermal, chemical, or microwave treatments are possible to improve anaerobic digestion (Hendriks and Zeeman G. 2009).

Rice straw is one of the most abundant and renewable energy sources in the world with approximately 916 million tons produced in 2009 (FAOSTAT, 2011). Rice straw is a fibrous, lignocellulosic biomass that makes it difficult to decompose but methane yields ranging from 92 to 280 l/kg of volatile solids have been achieved using pretreatment strategies (Mussoline et al., 2013).

Table 2. Cellulose, hemicellulose and lignin contents in common agricultural waste

Lignocellulosic waste	Cellulose (%)	Hemicellulose (%)	Lignin (%)
Grasses	25 - 40	35 - 50	10 – 30
Leaves	15 - 20	80 - 85	0
Nut shells	25 - 30	25 - 30	30 – 40
Maize stover	37.5 - 77	30	8.4 – 10.3
Rice straw	38.9 – 44.3	24 – 33.5	5.6 – 20.4
Rice husk	34.4 – 38.3	17.4 – 29.3	18.3 – 19.2
Bark sorghum	19.2	17.5	8.8
Fruit and vegetable waste	9	75	5
Sugarcane bagasse	45	20	30
Solid cattle manure	1.6 – 4.7	1.4 – 3.3	2.7 – 5.7
Sorted refuse	60	20	20

Adapted from Chandra et al., 2012, Billa et al., 1997, Raynal et al., 1995.

Fruits and Vegetables Waste Digestion

Fruit and vegetable waste are characterized by a high percentage of moisture (>80%), high organic content (volatile solids >95% of total solids) and are readily biodegraded, and therefore suits to energy recovery through anaerobic digestion (Appel et al., 2011). Fruit and vegetable waste are degraded very easily and are mostly co-digested with other feedstocks (Table 3).

Mata-Alvarez et al. (1992) reported on the performance of a mesophilic single-stage stirred digester for the treatment of a mixture of fruit and vegetable waste from a large food market, and found the maximum OLR that could be achieved was <3 g VS $.L^{-1}$ $.day^{-1}$. Carbohydrate-rich substrates such as fruit and vegetable waste are fast producers of volatile fatty acids leading to low pH and process inhibition (Mata-alvarez et al., 2000).

Vegetable waste could not be used as a sole substrate for anaerobic digestion without suitable trace elements supplementation. Digesters failed after approximately 3 HRT as a result of the synthesis of volatile fatty acids and reduction in pH. These changes could not be compensated on the long term by the use of chemical buffer solutions (Jiang et al., 2012).

Food Processing Residues Digestion

Food processing industries' effluents have very high concentrations in oil and grease, proteins and sugars and the oily wastewaters could be an abundant source of biomass to produce biogas. Lipids included in food waste and wastewater consists mainly of triacylglycerides and long-chain fatty acids. In anaerobic treatment systems, triacylglycerides can be hydrolyzed by extracellular lipases to long-chain fatty acids and glycerol. Glycerol is converted to acetate by the acidogenesis process while LCVFA are successively degraded via ß-oxidation pathway to acetate and hydrogen, which then are converted to methane (Figure 1). If compared with other organic matters like carbohydrates and proteins, lipids are attractive for biogas production due to the fact that they are reduced materials and have a high theoretical methane potential which can contribute to increase the biogas production. Lipid-rich waste such as oily wastewaters for example can be regarded as a large potential of renewable energy source. However, anaerobic treatment of organic waste with high lipid and protein content present several operational problems that limit the potential to obtain methane. The fats may solidify at lower temperatures and cause operational problems such as clogging and developing unpleasant odours. Pre-treatments are required to improve biodegradability and methane conversion of these kinds of substrates.

Manure digestion

The methane potential of manure comes from the digestion of the organic components present in the faeces and urines and also the straw if used as bedding material. Methane yields observed from anaerobic digestion of different agricultural waste are summarized in table 3. One can observe large variations for the same type of agricultural waste.

Table 3. Methane yield from anaerobic digestion of different agricultural waste

Type of agricultural waste	Methane $m^3.kg^{-1}$ VS
Fruit peels	0.24-0.52
Fruit pressings	0.28-0.5
Rotten fruits	0.21-0.51
Vegetable leaves	0.23-0.31
Straw (rice, maize,…)	0,092-0,33
Pig manure	0,36
Cow manure	0.28
Dairy cattle manure	0.15

Adapted from Nallathambi, 2004; Moller et al., 2004, Mussoline et al., 2013; Chandra et al., 2012.

One option to simplify by-product management and reduce the costs associated with biogas plant feedstock is to substitute dedicated crops with vegetal by-products. To this end, the potential biogas yield of agricultural by-products can be evaluated with batch trials under mesophilic conditions. Among considered biomasses (maize stalks, rice chaff, wheat straw, kiwi fruit, onions, expired dairy and dry bread mixture) the highest methane producer was the expired dairy product mixture, with a yield of 554 NL. $CH_4.kg^{-1}$ volatile solids (VS). Maize stalks and wheat straw produced the lowest yields, 214 and 285 NL. $CH_4.kg^{-1}$ VS, respectively (Menardo & Balsani, 2012).

Biogas production from co-digestion of whey and cow manure was enhanced with high biogas production (+21.8 %), with a good methane content (78.7%) in a biogas mixture as well as with the maximum removal efficiency (Hublin *et al.*, 2012).

The availability of trace metals as micro-nutrients plays a very significant role on the performance and stability of agricultural biogas digesters, which are operated with energy crops, animal excreta, crop residues, organic fraction of municipal solid waste or any other type of organic waste. Supplementation of essential trace elements has been shown to be important to maintain and even stimulate the digestion process in a number of studies (Osuna et al., 2003, Worm et al., 2009). This dependence can be explained by the fact that essential enzymes required for anaerobic biochemical reactions contain metals such as cobalt, nickel, iron, zinc, molybdenum and tungsten, in the form of coenzymes and cofactors (Fermoso et al., 2009).

The use of a trace element solution according to Gonzalez-Gil et al. (2001) allowed stable digestion at a low loading of 2 g $VS.L^{-1}$ $.day^{-1}$ but the recommended dose may have been insufficient to satisfy trace element requirements when higher loadings were applied.

The unavailability of these elements in biogas digesters is probably the first reason of poor process efficiency without any other obvious reason, despite proper management and control of other operational and environmental parameters. However, trace metals requirements of biogas digesters operated with solid biomass are not often reported in literature (Demirel and Scherer, 2011).

Mixing is also an essential parameter for biogas production. Too much mixing stresses the microorganisms and mixing foam or crust layer occurs. Methane-forming microorganisms grow slowly, with a doubling time of around 5–16 days. Therefore, the hydraulic retention time should be at least 10–15 days (Rajendran et al., 2012).

Digesters Design Used for Agricultural Waste: State of the Art

There are several main types of anaerobic digesters, covered lagoons, complete mix, plug flow and fixed film. There are also a number of hybrid systems combining diverse types because no single system is the best in most situations. The digester can be wet or dry, mesophilic or thermophilic, and single or multistage.

Complicated construction, difficult operations of the systems, high investment, and maintenance costs have pushed farmers to adopt cheaper and simpler anaerobic systems (Georgakakis, 2003).

For agricultural anaerobic digestion plants, robustness and simplicity are amongst the main parameters. Farmers will never accept and use sophisticated systems that require expert audit and frequent consultancy by technicians. This rusticity and low cost remain true in developing countries rural regions, whereas in developed countries, high technology reactors, often monitored on distance by the manufacturer have been introduced.

It has to be noticed that at this rustic, rural and small scale, household waste are always mixed together with cattle or crops wastes: the objective there is to treat ANY kind of organic waste. As already mentioned, anaerobic digestion used to exist for millenniums, so that long time before any bioreactor modelling could help for a comprehensive approach of the complex phenomena involved. Thus, very rustic reactors already existed and still exist in developing countries particularly.

Depending on different parameters, such reactors involved may be categorized in different ways. First, depending on the aspect of wastes, i.e. solid or liquid ones. Solid may be manure when bovine or horse manure is mixed with straw, rushes, cereals, corn stalks, soya cakes. Liquid may be swine manure, whey...The design of the reactor in this case is mainly concerned by the introduction vessel and apparatus and as well as the power involved in mixing.

Then, the operating conditions depend on the temperature range chosen whether they are in psychro, meso or thermophilic range. As far as the degradation kinetic is mainly influenced by the operation temperature, as far as it has obviously a consequence on the residence time of waste inside the reactor, so the volume of this later, few biogas reactors are driven under psychrophilic conditions, which would lead to an unacceptably large reactor. On the other side, running a reactor under thermophilic conditions will consume a lot of energy and/or a large amount of insulation for thermal energy to be kept in. Therefore, thermophilic conditions could be used only if the effluent is hot, e.g. process water from dairy industry. Finally, a huge majority of such bioreactors at farms are run under mesophilic conditions, heated or not depending on the local climate conditions and the more or less rusticity of the apparatus.

The next difference is either one want to keep the technology simple, or want to improve the energy production and/or the rate of waste treatment.

The efficiency may be improved by giving to the microbial communities their best growing conditions, particularly for the pH parameter. Thus, this will lead to a 2 step reactor, including an over-cost for pipes, pumps, civil engineering that may be covered by the increase in the biogas rate production.

Back to historical vessels, these are still in operation in many developing countries such as India, Pakistan, China or African countries. In these countries, not only manure and farm residues are digested, but also cooking residues and human droppings. As many of these countries are in tropical regions, no extra heating is needed to maintain mesophilic conditions, leading to a rustic design. In addition, as far as the reactor remain rather small as it is at a family scale, it remain quite easy and inexpensive to excavate in order to build the digester underground, and thus, take advantage of a cheap isolation: the soil.

The traditional model of this kind of digester is known as "chinese model" made of masonry, with crude bricks covered of cement or a kind of plaster. It is a fixed dome model, usually built underground. The size varies between 1 and 150 m^3. Some modifications have been made such as a wall built in the middle of the hemispherical shape or a reduced diameter of the arch for the Deenbandhu model (Rajendran *et al.*,2012).The necessary minimum of pressure for the biogas is given by the production of the biogas and by the introduction of the manure. The ejection of the digested products is made by hydraulic flushing, the introduction pipe being on an opposite diameter.

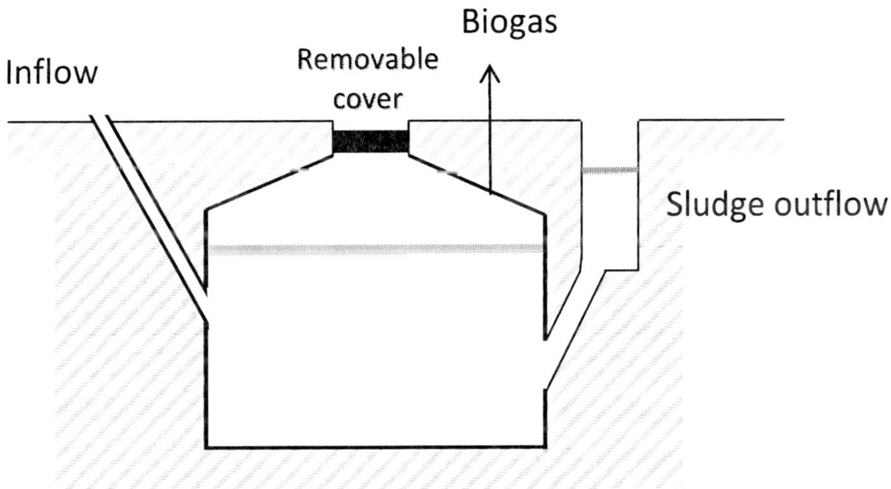

Figure 2. Chinese's traditional farm waste digester.

The rural digester used in India is of a different design where a floating drum digester model was developed in the sixties. This floating drum digester produces biogas at a constant pressure with variable volume. For small-medium size farms the size varies from 5 to 15 m^3. This design is much closed to the ancient huge gasometers used during the nineteenth century in western large cities. It consists of two sliding metallic cylinders, the upper one moving along a guide leading to a so-called floating dome. Thus, the necessary pressure is given by the weight of this dome onto the biogas layer.

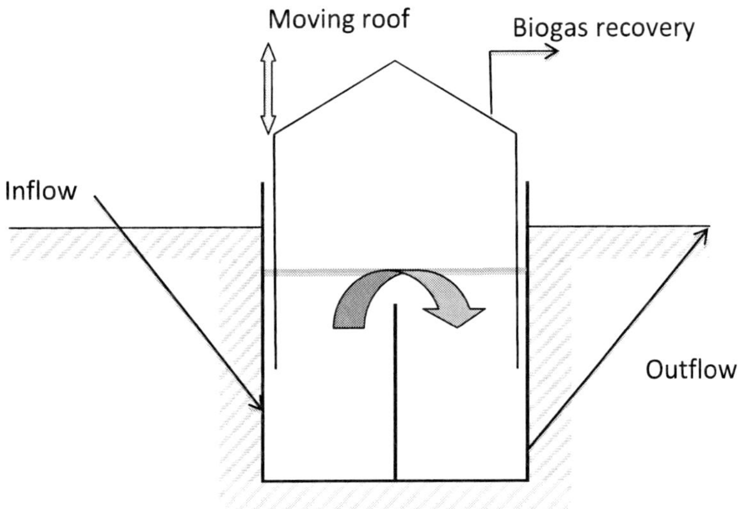

Figure 3. Indian rural digester.

In western countries, digesters are in a larger scale, industrialized, and heated. Few original technologies are used as the objective remain to keep the device rustic, thus robust and no need to be ran by specialists, but generally by the farmer himself. For example, while there are many cases of two-stage reactors (hydrolysis/acidogenesis then methanogenesis) for the treatment of solid household waste by municipalities, they are not found in the agricultural sector. Thereby, the differences mainly come from the design of the agitation and/or gas extraction device.

First, one can meet reactors mixed by the biogas itself, or even not mixed at all. In this case the digester is made of two chambers. A valve links the gaseous skies of the two chambers. The outlet of biogas is in chamber 2. The production of biogas in chamber 1 pushes the waste in chamber 2 that purges the biogas. Before the liquid level reaches the outlet, the valve is suddenly opened, which balances the pressure between the two gaseous skies, so that a wave movement occurs mixing the liquid and so the liquid level in chamber 1 rises and purges the biogas in this chamber. It is easy to understand that this kind of reactor is only used for liquid effluents.

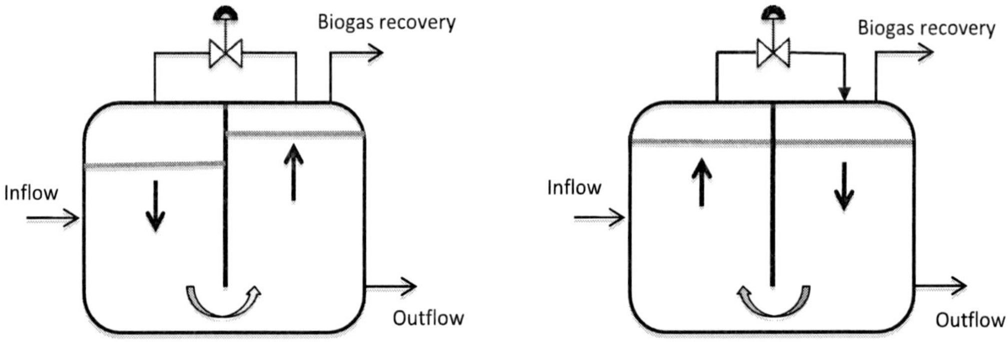

Figure 4. So called "pulsed" reactor.

In sequencing batch reactor (SBR) the problem is even simpler, because there is no agitation at all. It looks like a glow tube, a kind of rubberized cover or tarpaulin. Waste are introduced on a daily base, biogas purged continuously or sequentially, same for the digestate. These fermenters are only used on a very small scale.

Then, two kinds of ideal reactors may be used: continuous stirred tank reactor (CSTR), plug-flow reactor (PFR).

Plug-flow digesters are unmixed systems in which waste flows semi-continuously as a plug through a horizontal reactor. The reactor may be an in-ground tubular tank or a covered, concrete-lined trench. This design typically handles a solid content of 10-14% and operates at mesophilic temperatures. HRTs vary from 20 to 30 days. The plug-flow digester is suitable for scraped-manure operations. Fibrous solids are removed for post digestion.

Plug-flow digesters with periodic loading mechanism are more adapted for the majority of small or medium scale farms according to the costs and operational complexities than fully mechanized digestion plants. A dual-compartment plug-flow reactor equipped with a passive heating system was designed and experimentally operated by purpose of demonstrating a simple and low cost technology for handling the biodegradable agricultural waste. The reactor was successfully started up with pig faeces as feedstock under a quasi-continuous loading and semi-dry condition with average total solids content of 12.8% inside the digester and an average organic loading rate of 2.06kg VS $.m^{-3}.d^{-1}$. The start-up phase was followed by co-digestion of pig faeces and pre-treated cotton stalk. The biogas and methane yield were measured for single digestion stage as 0.332 and 0.202 $m^3.kg^{-1}$ a VS and for co-digestion stage as 0.482 and 0.325 $m^3.kg^{-1}$ VS respectively (Adl et al., 2012).

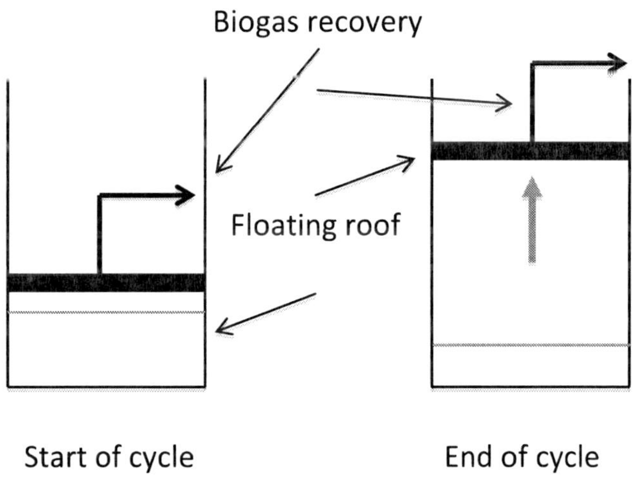

Figure 5. Sequencing Batch Reactor model (SBR).

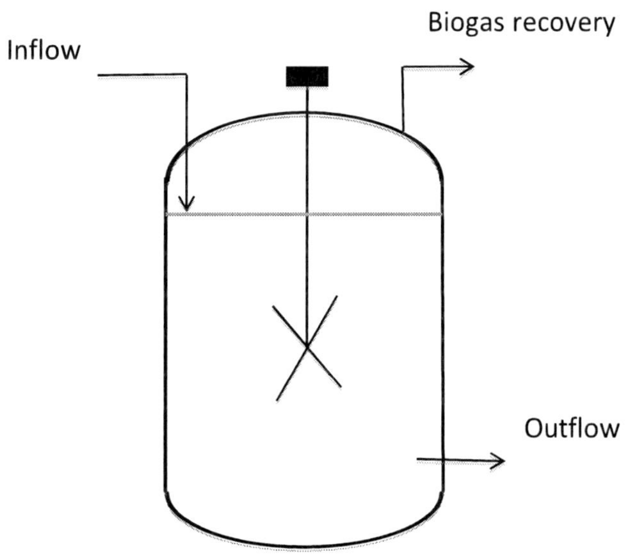

Figure 6. Continuous Stirred tank reactor model (CSTR).

Most of the digesters used in farm applications to treat agricultural waste are based on the first ideal model. In this model, the reactor is assumed to be well mixed so that the contents are uniform in composition at any time. In fact, agricultural reactors are a little far from ideal because the introduction of waste is rarely continuous, as far as solid wastes are introduced sequentially. But the frequency being of some times per hour, on a large time scale, we can consider this introduction as continuous. But theoretically we should call them "semi-continuous", or "pseudo-continuous".

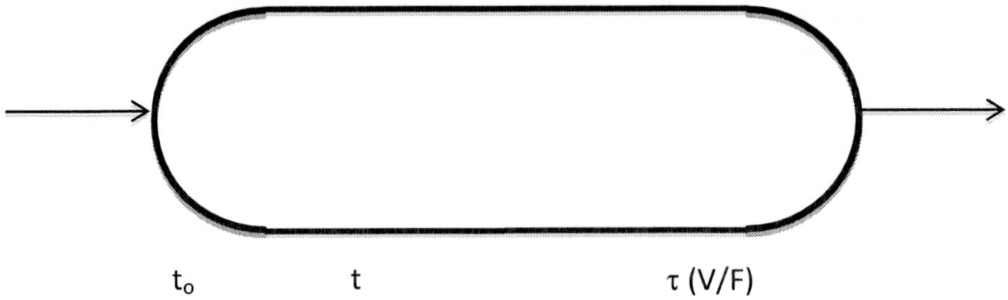

Figure 7. Plug-Flow Reactor model (PFR).

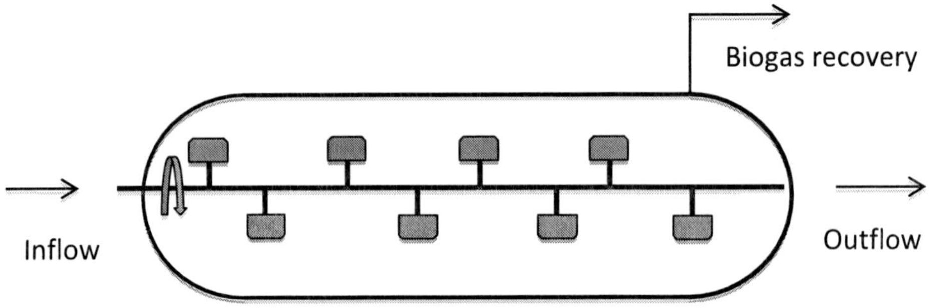

Figure 8. Near plug-flow real fermenter.

Then two kinds of apparatus may be distinguished depending on the biogas storage site. The simplest reactor used only one vessel, with a flexible membrane as a roof.

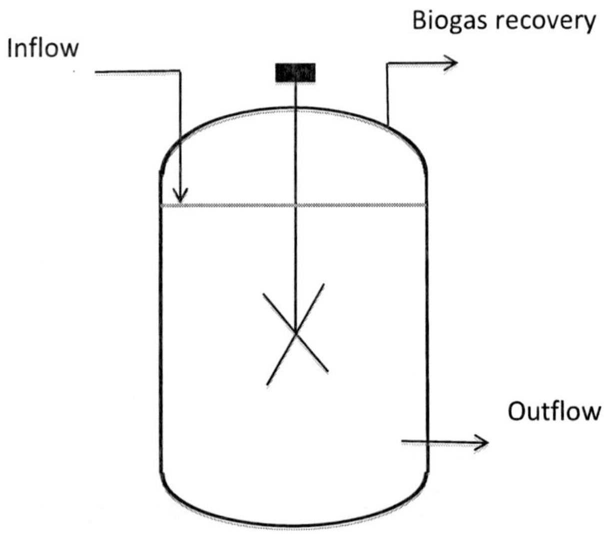

Figure 9. Classical single chamber biogas reactor.

Single membrane roofs may be the source of problems. In fact, any biogas reactor is equipped with a discharge valve and a flare. Single roof membrane is often black, and thus, accumulates energy from solar radiation. Thus, by applying the ideal gas law, the pressure increases and the gas is released by the valve. However in winter, if snow accumulates on the dome, the pressure increases, so the gas is released, there is less resistance and the dome collapses like a donut, and this vicious circle may lead to fatal damages to the membrane.

The most classical model uses two vessels: the reactor itself has a rigid ceiling with a little sky for gas production. For instance in a 6 m high fermenter, the interface is maintained almost at 5.5 m from the bottom. The biogas is extracted to the biogas storage tank. This one is both used for biogas and digestate storage. Digestate is extracted from the main reactor simply by hydraulic flush from the bottom of the fermenter as far as solid wastes are mechanically introduced through the top. As far as the storage tank is used for the digestate storage, the digestion is still running, for large molecules needing a long residence time. This is why this storage tank is usually called post-digester. In such apparatus, classical values for residence time are 30 to 60 days in the main stirred reactor, 6 month in the post digester. The other advantage is that by storing the biogas, this one is mixed and variations in composition are smoothed so that the exploitation of the biogas is easier and more constant particularly in CHP appliance. This type of reactor is design for any size from some hundreds cubic meter to some tens of thousands. Major constructors use this design (such as MT-Energie, EnviTec Biogas AG, Biogas Nord AG, OWS, WELtec GmbHBioPower, B.T.S.®Biogas, BioConstruct G,mbH, InSource Energy Ltd, BIGADAN A/G, NASKEO...) particularly in Germany. In order to avoid the problems mentioned above about single roof technology, in most of cases, the biogas storage tank is covered by a twin membrane: the moving one directly in contact with the biogas, protected from the outside by a second one, supported by a light frame, or simply held in tension by a small compressor that blows air continuously between the two membranes.

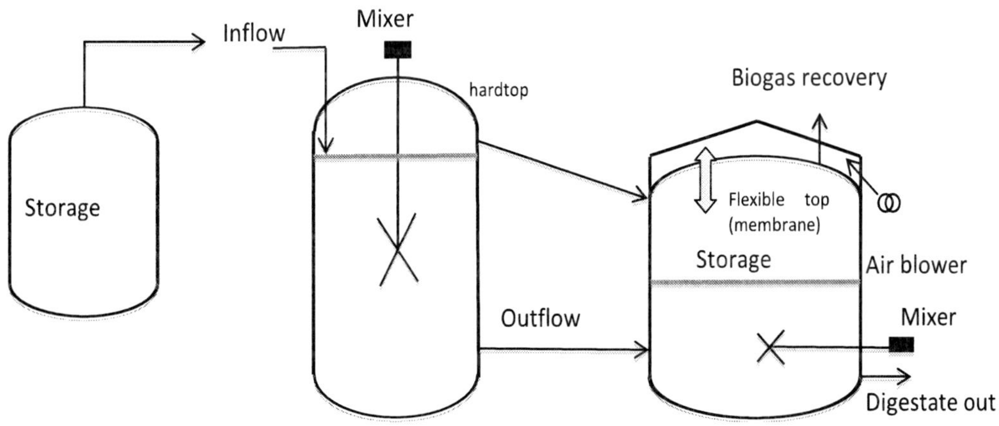

Figure 10. Two stages traditional agricultural waste's biogas reactor.

This is the model that has been designed mostly in Germany with the development of the so-called "energetic farms". In this extreme case, it is not question of turning *waste* into a resource, but to produce organic matter specifically grown as an energy resource, i.e. corn silage. Thanks to this voluntary policy with attractive prices for the electricity, the market for large-scale fermenters has exploded, leading to more that 6 800 plants, (amongst those 6 000 "farm" digesters) with an electric power of almost 2.7 GW and an annual production of more that 16 000 GWh and, for instance, in Güstrow, EnviTec which built the largest biogas plant in the world, whose poduction is amongst 104 m^3/hr. (EUROBSERV'ER, 2012).

Some nuances may be introduced by the stirring technology used. Two kinds are developed. Horizontal slow large mixers fixed at the bottom of the fermenter, like the slurry plowing arms or trussed racks used in wastewater sedimentation tanks. These always turn. In fact, problems may occur in case of breakdown, where the fermenter has to be fully emptied. The other technology used is an immersed lateral stirring device, of which angle may be adjusted depending on the dry matter content of the introduced slurry. These are fixed on a rack and may be removed from the fermenter for maintenance and repair. The same in a smaller size are used in the post-digester to avoid sedimentation and help the gas release. Large diameter reactors are used for multiple reasons: geotechnical (for a better distribution of the load on the ground), visual impact (no high towers), security and safety for maintenance (roof easily accessible), better extraction of the biogas (less pressure).

Plug-flow reactors are rarely used for agricultural waste. PFR principle means that a waste slice will move independently from the others and being transformed depending on the residence time. This kind has been mainly designed to be portable. These digesters have a constant volume under variable pressure during biogas production. The size of such digesters varies from 2.4 to 7.5 m^3. Plug flow designs are suitable for manure, and operate semi-continuously with a HRT between 20 and 30 days, and with solid contents varying between 11 and 14% (Rajendran *et al.*, 2012).

Lagoons with a floating cover could also be used as a digester, which is very cheap for farmers. This design typically handles a solids content of < 2% and operates at ambient temperatures. Hydraulic retention times (HRT) vary from 35 days in the south to 60 days in more northerly regions. The covered lagoon digester is suitable for flushed-manure operations. Fibrous solids are removed for pre-digestion. Biogas production tends to vary seasonally due to temperature fluctuations (Wilkie, 2005).

Then remain the technology used for the reactor heating. But the technology used is not only a technological problem, but an economic or regulatory issue. Before heating, insulation has to be improved. In hilly regions, the fermenters may be totally or partly underground for a better insulation, but main of the industrial realization are built above the ground on a concrete slab. In fact, in France for instance, where the energy recovery is performed through a CHP, to get the best price for the electricity, heat energy must be exploited. So a part of this form of energy is used to heat the fermenter. A spiral tube stuck along the wall of the fermenter is the transfer medium generally used. Another part may be used to preheat the liquid manure in winter through heat exchangers to limit the temperature drop.

As already mentioned, in tropical countries for rustic fermenters, these remain unheated. There remain the countries where heat energy is not systematically exploited: when biogas is transformed elsewhere or transformed in pure methane to be injected in the natural gas network, e.g. Germany, or where there is another easiest source of energy: solar energy, Greece, Italy, Spain... Some have developed solar heated fermenters. (Axaopoulos (2), 2001, 2003, Yettou, 2007).

Household Digester and Agricultural Waste

Rajendran *et al.*, (2012) have made a review of different substrates with biogas yield used in household biogas digesters (Table 4). Cattle manure was a traditional source for biogas production in the past. The methane content was high in pig manure, around 60%, followed by cow dung with 50%. Crop residues are some underexploited substrates for the domestic biogas production.

Table 4. Different substrates and biogas yield for household anaerobic digesters

Substrate	Substrate classification	Dry matter (%)	Biogas yield
Agricultural residues	Rice straw	91	$0.55 – 0.62\ m^3\ kg\ TS^{-1}$
	Wheat straw	91	$0.188\ m^3\ kg\ VS^{-1}$
	Maize straw	86	$0.4 – 1\ m^3\ kg\ TS^{-1}$
	Grass	88	$0.28 – 0.55\ m^3\ kg\ VS^{-1}$
	Coffee pulp	28	$0.3 – 0.45\ m^3\ kg\ VS^{-1}$
	Corn stalk	80	$0.35 – 0.48\ m^3\ kg\ TS^{-1}$
Manure	Cow	38	$0.6 – 0.8\ m^3\ kg\ TS^{-1}$
	Pig	20 – 25	$0.27- 0.45\ m^3\ kg\ TS^{-1}$
	Poultry	89	$0.3 – 0.8\ m^3\ kg\ TS^{-1}$
	Horse	28	$0.4 – 0.6\ m^3\ kg\ TS^{-1}$
Food waste	Vegetable waste	5 –25	$0.4\ m^3\ kg\ TS^{-1}$
	Egg waste	25	$0.97\ m^3\ kg\ TS^{-1}$
	Cereals	85 - 90	$0.4 – 0.9\ m^3\ kg\ TS^{-1}$

Adapted from Rajendran et al., 2012.

RESEARCH TRENDS

Although anaerobic digestion is a mature technology, the process is not yet fully understood and optimization of the current technology is still needed. There is no standard digester for different reasons: in one hand, the situations are very diverse in operating conditions and in the other one, there remain many research issues still incomplete concerning microbiology, microbial ecology, metabolic pathways, modelling, technology… Various fields can be identified in which more research is necessary to further optimize anaerobic digestion processes:

a) Optimization of anaerobic digestion
   - mproved knowledge and optimization of the ecosystem in case of codigestion
   - Optimization of operating units

b) Improvement of processes and appearance of new technologies
   - Rusticity
   - Lower investment and operation costs

   - Techniques for effective purification of biogas, more rustic and reduced cost
   - Monitoring / plant automation

In particular, we can learn from termite's rumen models where back and forth cycles of mechanical/chemical/enzymatic pre-treatment and microbial fermentation (rumination) occur.

For example it could be possible to bio-increase the efficiency with protozoa and fungi (cellulases, xylanases and lignases producers) found in gastrointestinal tract or to create microaerobic conditions to stimulate added or indigenous hydrolytic aerobic or facultative microorganisms.

Pre-Treatment for Anaerobic Digestion Enhancement

In practice, degradation efficiency hardly exceeds 60%, with Y_{CH4} average 0.3 Nm^3/kg VS fed (except some residues such as agro-food or slaughterhouse > 80%) and the limiting step remain hydrolysis, the first step of anaerobic digestion (Guiot et al., 2009). To enhance hydrolysis, typically hydrolytic pre-treatment are employed to optimize the biomass-to-energy conversion. The goal is to solubilise the organic matter in order to turn the organic compounds more biodegradable. The use of pre-treatment methods is particularly useful when digesting lignocellulosic biomass and wooden fractions (Ward et al., 2008). The lignin, hemicellulotic and cellulotic structures are hardly biodegradable by the anaerobic microorganisms, resulting in very long retention times and a low digestion output. Various pre-treatment techniques have been developed these recent years: chemical (acid or alkaline hydrolysis, ozonation); thermal or thermo-chemical (steam or wet explosion, microwave ...); physical (high pressure homogenizers, impact grinding stirred ball, mills ultrasonic homogenizers, high performance pulses, lysate centrifuge, gamma irradiation) and biological (enzymes) treatment. With summer switchgrass, use of thermo-chemical pre-treatment increases the methane yield from 100 Nm^3 $CH_4.T^{-1}$ VS to 260 Nm^3 $CH_4.T^{-1}$ (Guiot et al., 2009). Potential methane production of pre-treated straw increased from 2 to 29% from the untreated sample value (Ribeiro et al., 2012). The higher enhancement was observed for thermo-chemical techniques coupled with acid or alkaline hydrolysis (microwave + HCl or NaOH). The lower enhancement was observed for alkaline hydrolysis. Pre-treatment of straw reduced from 18 to 30% the time needed to obtain 80% of the maximum methane production and enhanced kinetic methane production from 13 to 40%. Pre-treatments increase cellulose and hemicellulose degradability by structural changes enhancing the accessibility to micro-organisms and enzymes (Figure 11).

Hydrothermal pre-treatment (water under high pressure and temperature) of lignocellulosic biomass for enhanced biogas production is gaining high importance. The major advantages are no addition of chemicals and no need for corrosion-resistant materials for hydrolysis reactors in this process. Hydrothermal pre-treatment enhances the accessible and susceptible surface area of the cellulose and make it more accessible to hydrolytic enzymes (Taherzadeh and Karimi, 2008). Addition of alkaline compounds (like NaOH) is essential because hydrothermal pre-treatment decreases the pH of biomass slurry, which might be caused by hydrothermal oxidation of sulphur and phosphorus present in the biomass, which results into production of inorganic acids (Chandra et al., 2012a). Rice straw hydrothermal pre-treated substrate (200◦C and 10 minutes residence time) followed by addition of 5% NaOH has resulted into biogas and methane production yields of 315.9 L/kg VS and 132.7 L/kg VS, respectively. The same has resulted into an increase of 225.6% in biogas production and 222.0% in methane production relative to untreated rice straw substrate (Chandra et al. 2012b)

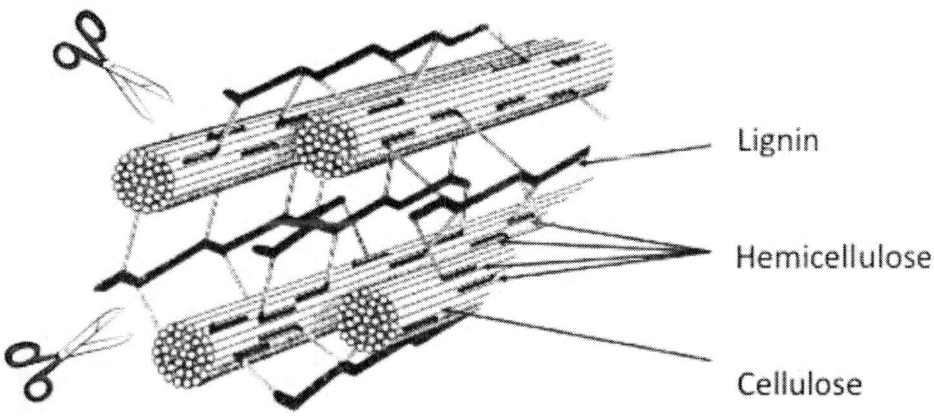

Figure 11. Pretreatment action on wood structure of straw.

Operation Monitoring

Anaerobic digestion plants are commonly regulated based on at-line or off-line analytical results. Few process parameters are routinely recorded on-line and they are limited to accessible data such as pH, redox potential, gas production rate, and flow rates. The technological development within the field of process monitoring in the established fermentation industry is attractive to analyse as it has many similarities with the anaerobic digestion. Available techniques can be organized in four main groups (Table 5).

Table 5. Available techniques for parameters of anaerobic digestion process

Chromatographic	Electro-chemical	Spectroscopic	Diverse
Gas Chromatography (GC) GC-Headspace High Pressure Liquid Chromatography (HPLC)	pH Redox potential Electronic sensors (tongue, nose)	Visual (VIS) Infrared (IR) Ultraviolet (UV) Fluorescence (FLU) Raman	Titration Mass Spectrometry (MS) Microwaves Acoustic chemometrics

Adapted from Madsen et al., 2011.

Bioaugmentation or Enzyme Addition

Enzymes could be added into a single-stage anaerobic digester or could be used to pre-treat the biomass material prior to anaerobic digestion. In two-stages anaerobic digestion, enzymes could be added to the hydrolysis stage prior to biogasification.

Biological degradation of lignocellulosic material is facilitated by enzymes, such as cellulase and hemicellulase, which are produced by the microorganisms in the digester. However, enzymatic hydrolysis of lignocelluloses with no pre-treatment is usually not so effective because of the high stability of the materials against enzymatic attacks (Parawira, 2012). The crystallinity of cellulose, its accessible surface area and protection by lignin and hemicelluloses, degree of cellulose polymerization, and degree of acetylation of hemicelluloses are the main factors considered as affecting the rate of biological pre-treatment of lignocelluloses by enzymes. Use of vital microorganisms seems more dynamic and efficient than free enzymes due to their ability for regeneration and concomitant production of diverse enzyme responding to the given substrate (Parawira, 2012). After biological lignin removal the straw cellulose was more accessible for anaerobic digestion. Biological ensilage additives with hetero- and homo-fermentative activity (Silasil energy®) as well as enzymes (Sil-all 4 x4®) or bacteria and yeasts (Microferm®) effectively increased the biogas production on organic dry matter of maize (11.8, 10.1, and 14.7%, respectively) after 7 weeks of pre-treatment (Vervaeren et al., 2010 cited by Parawira, 2012). Bagi et al. (2007) shown an increase of biogas production by about 60–70% due to inoculation of biogas reactors of up to 5m^3 size with external hydrogen-producing bacteria with cellulolytic activity (*Caldicellulosyruptor saccharolyticus*). There is need for further research on the effect of adding enzymes to anaerobic reactors digesting lignocellulosic material, given its potential as a source of raw material for biogas production (Parawira, 2012). Enzymes or lipolytic organisms can be used to improve biodegradation of rich-lipids waste. Adding enzymes into anaerobic digesters treating food-processing waste have been reported to result in improved digestion and biogas production. Lipases have shown to be a very promising alternative for degrading rich-lipids wastewaters generated by dairy and slaughterhouses industries (Mendes et al., 2006).

Due to the high cost of commercial enzymes, the use of hybrid technology (enzymatic treatment combined with biological treatment) using enzymatic pools produced through solid-state fermentation from low-cost industrial waste may represent an important step in the treatment of wastewaters with high oil and grease contents. An application of low-cost enzymes for pre-treatment of effluents from poultry slaughterhouses was reported by Valladao et al. (2007). Enhanced anaerobic treatment efficiency relative to raw effluent (COD removal efficiency of 85% versus 53%) was achieved when a 0.1% concentration of enzymatic pool (hydrolases produced from solid-state fermentation) was used in the pretreatment stage with 1200 mg oil and grease.l^{-1}.

Electromethanogenesis and Bioelectrochemical Reactor

Among the new trends in research should be mentioned the development of electromethanogenesis. In this case, microaerobic conditions are controlled by water electrolysis. Methane production increases by 10-40% depending on applied power and organic load (Guiot *et al.*, 2009). Bioelectrochemical reactor (BER) can stimulate microbial metabolism and affect the growth rate of microorganisms by controlling the electron flow in the culture medium (Thrash et al., 2007). Bacteria on the anode of a microbial fuel cell (MFC) break down organic matter to produce electrons and protons ("electrolyzing" organic matter) into electrons, protons, and carbon dioxide. The conversion of acetate or hydrogen into methane is a thermodynamically favourable reaction. Both acetate and hydrogen are used by methanogenic microorganisms to produce methane. It was discovered that methanogens can directly accept electrons from an electrode (and, thus, presumably from other microorganisms), so that an organic fuel (e.g., acetic acid) or hydrogen gas are no more compulsory to produce methane. It was shown that a methanogenic biofilm on a cathode of a microbial electrolysis cell (MEC) could produce methane at much greater rates than could be explained by possible rates of hydrogen formation at the cathode (Cheng et al., 2009). Carbon dioxide is used by methanogens in a MEC to create methane. This ability of methanogens to turn electrical current into methane means that renewable sources of electricity, for example from a windmill or solar panel, could be converted into methane (Logan, 2009). Electromethanogenesis can be used to convert electrical current produced from renewable energy sources (such as wind, solar, or biomass) into a biofuel (methane) as well as serving as a method for the capture of carbon dioxide (Cheng et al., 2009). The production of methane using MECs is at a very early stage and further developments are needed to achieve high methane production rates. The reactor design is one of the key factors for electricity production efficiency. One approach consists in observing and copying the nature's great success for minimization of resistances and maximization of mass and energy transfers. The constructal approach can be used in a feed-through MFC as a tool to improve mass and energy transfers and to decrease the pressure loss between input and output of the system (Lepage et al., 2011).

Till now, there is no full-scale application of electromethanogenesis but only lab-scale experiments. A cylindrical bioelectrochemical reactor (BER) containing carbon fibre textiles (CFT; BER + CFT) with bioelectrochemical characteristics and packed-bed systems have been studied for degradation of a garbage slurry and recovery of biogas by applying 10% of dog food slurry (Sasaki et al., 2013). The working electrode potential was electrochemically regulated at −0.8 V (vs. Ag/AgCl). Stable methane production of 9.37 L-CH_4.L^{-1}.day^{-1} and chemical oxygen demand (COD) removal of 62.5% were observed, even at a high organic loading rate (OLR) of 89.3 g-COD.L^{-1} day^{-1}. Given energy as methane (372.6 kJ.L^{-1} day^{-1}) was much higher than input electric energy to the working electrode (0.6 kJ.L^{-1} day^{-1}) at this OLR. Methanogens were highly retained in CFT by direct attachment to the cathodic working electrodes (52.3%; ratio of methanogens to prokaryotes), compared with the suspended fraction (31.2%), probably contributing to the acceleration of organic material degradation and removal of organic acids.

New Reactor Design for Agricultural Waste

Digestion of *Jatropha curcas* seed cake was investigated in two-stages (acidogenic and methanogenic) anaerobic bioreactors without pH adjustment. Acidogenic reactors were fed once daily with a slurry of 1:10 *Jatropha curcas* seed cake:water containing approximately 100 g of chemical oxygen demand/L. Organic loading rates were 2.5, 3.3, 5, 10, and 20 kg COD.m^{-3}.d^{-1}, which corresponded to hydraulic retention times of 40, 30, 20, 10, and 5 days, respectively, for each reactor stage. The maximum methane yield (340 L at STP/kg of chemical oxygen demand degraded) was observed at an organic loading rate of 3.3 kg COD.m^{-3}.d^{-1} (hydraulic retention times = 30 days for each stage). At this organic loading rate and hydraulic retention time, the chemical oxygen demand degradation efficiency was 65%. The average pH in the acidogenic and methanogenic reactors was 4.9 and 7.4, respectively. This study demonstrates high methane yield and degradation extent of *Jatropha curcas* seed cake in a two-stages anaerobic process without chemical addition for pH adjustment (Sinbuathong et al., 2012).

Anaerobic Moving Biofilm Bed Reactor (AMBBR): A Fixed-Film Digester

A fixed-film digester immobilizes bacteria on a packing material, or media, within the reactor vessel, thereby preventing washout of microbial biomass. This design typically handles a solids content < 2% and can operate at ambient (15-20°C) or higher temperatures. At ambient temperature, the lower rate of metabolism is offset by a high microbial mass. HRT is in the order of 2-4 days. The fixed-film digester is suitable for flushed-manure operations. Fibrous solids are removed through pre-digestion. Higher levels of solids can be tolerated, depending on biodegradability. Also, the fixed-film design can tolerate the presence of some fine sand grains in the wastewater without impacting performances.

An anaerobic moving bed-biofilm reactor (AMBBR) like depicted in figure 12 was developed to treat dairy effluent including whey (Wang et al., 2009). The moving support had a density of 0.95 (near 1 with biofilm). High strength milk permeate derived from ultra-filtration based cheese making process was treated in the AMBBR under mesophilic (35°C) condition. Total chemical oxygen demand (TCOD) removal efficiencies of 86.3–73.2% were achieved at organic loading rates (OLR) of 2.0–20.0 g $TCOD.L^{-1}.d^{-1}$. A mass balance model gave values of methane yield coefficient ($Y_{G/S}$) and cell maintenance coefficient (k_m) of 0.341 L $CH_4.g^{-1}$ $TCOD_{removed}$ and 0.1808 g $TCOD_{removed}.g^{-1}$ $VSS.d^{-1}$, respectively. The maximum substrate utilization rate U_{max} was determined as 89.3 g $TCOD.L^{-1}.d^{-1}$. Volumetric methane production rates (VMPR) were shown to correlate with the biodegradable TCOD concentration. The AMBBR is simple to built and simple to manage. The mixing can be realized with an impeller.

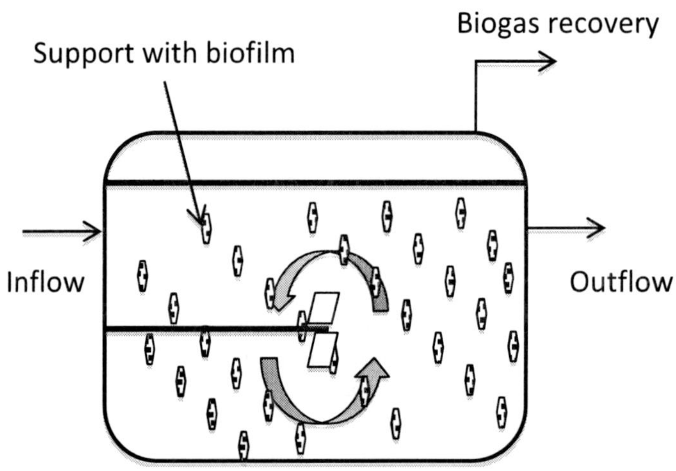

Figure 12. Diagram of an anaerobic moving biofilm bed reactor (AMBBR).

Codigestion

Codigestion consists in the digestion of a mixture of two or more substrates with complementary characteristics so that to enhance biogas production. In the area of agricultural waste, first reports seem to be dated in the late seventies (Hill, 1983). The main issue for co-digestion process consists in balancing C/N ratio, but also other several parameters in the co-substrate mixture such as macro- and micronutrients, pH, inhibitors/toxic compounds, biodegradable organic matter and dry matter (Hartmann and Ahring, 2005; Mata-Alvarez et al., 2011). Ideal co-substrates for manures are agricultural waste as well as crops, with a lower alkalinity and a higher C/N ratio. One of the pioneer countries in Europe applying co-digestion at farm level was Denmark, treating around 1.5 million tons.year^{-1} of manures mixed with another organic waste and preferably at the thermophilic range (Nielsen and Angelidaki, 2008). Comparing the single waste digestions with co-digestion of combined ones in the case of co-digestion of cattle manure (solid and liquid) with maize and fruit-processing waste, biogas production increased from 0.45 to 0.62 m^3.kg^{-1} VS operating at proper thermophilic conditions (55°C instead of 47°C) and methane content was also higher: from 52% to 61% (Cavinato et al., 2010).

Table 6. Waste introduced in a pilot digester in a French agricultural college (personal communication)

	Waste type	T/y	
	liquid cattle manure	600	
Agricultural waste	cattle manure (with straw)	860	
	green waters	360	
	Subtotal		1820
	grass clippings	5	
	hepatopancreas from snail processing	5	
	whey	250	
	cereal processing residues	200	
other wastes	mushroom production waste	30	
	food grease	10	
	stomach contents from abattoirs	180	
	Canteen waste	14	
	Apple juice process residues	200	
	Subtotal		894
	TOTAL		**2714**

Very little research have been conducted on digestion of multi-component substrates (Misi and Froster, 2001; Alvarez and Liden, 2008) to improve methane production. A study investigated the possibilities of improving methane yield from anaerobic digestion of multi-component (cow manure and mixtures of biodegradable agro and household waste) in bench scale laboratory reactors and in full scale reactors supplying gas to one household under local conditions in Bangladesh (Ashekuzzaman and Poulsen, 2011). Results from both bench and full-scale experiments fitted well and showed that using multi-component substrates increases the methane yield much more than what would be expected from digestion of single substrates (28-177% higher).

To balance the ration introduced, agricultural waste may be mixed wit some others. As an example, in table 6 is shown the ration used by a roughly small size (450 m^3, 45 kW CHP electric power) used in a French agricultural high school. (Lycée Reinach, 73290 La Motte Servolex).

CONCLUSION

Anaerobic digestion is an efficient way to turn organic agricultural waste into biogas and renewable energy. It is a fancy way both:

- To treat these organic waste (to contribute to the global waste treatment)
- To limit the free emission of CH_4 in the atmosphere by controlling the fermentation process (to contribute to the limitation of the greenhouse effect)
- To produce a renewable energy (to contribute to the energy transition)
- To produce this energy locally, everywhere on the planet where there is agriculture (to contribute to the energy independence, particularly for islands, isolated communities…)
- To participate to the farmers' income (to contribute to maintain a local and small scale agriculture).

New regulations introduced in European countries about the organic waste from canteens and restaurants and any food handcraft are new opportunities, where the rations for the digesters could be balanced and the yields improved, e.g. in Japan, a typical biogas plant in successful operation treats livestock wastes (25,171 tons) collected from farms and food wastes collected from food factories (4,827 tons) (Li and Kobayashi, 2010). Animal and food wastes are mixed in a storage tank before being fed to a 2,100 m^3 mesophilic digester or a 600 m^3 thermophilic digester with HRT and treatment capacities equal respectively to 33 d and 62.8 $T.d^{-1}$ and 26 d and 23.2 $T.d^{-1}$. The volumetric biogas production rate varies from 0.9 $Nm^3.d^{-1}.m^{-3}$ for the mesophilic digester and 1 $Nm^3.d^{-1}.m^{-3}$ for the thermophilic digester. Methane concentration was 55.4% for both digesters.

The growing market of biomethane also opens new possibilities where there is no local use for electricity and/or for the wasted heat from CHP.

Progresses have to be made in reactor's design, but designers and researchers should never forget that rusticity, robustness, liability and facility for running and maintenance are amongst the major qualities requested if we want to see farmers accept to appropriate this technology and take benefits of it.

REFERENCES

Alvarez, R. & Liden, G., (2008). Semi-continuous co-digestion of solid slaughterhouse waste, manure, and fruit and vegetable waste. *Renew. Energy*, 33, 726–734.

Appels, L.; Lauwers, J.; Degrève, J.; Helsen, L.; Lievens, B. & Willems, K. (2011). Anaerobic digestion in global bio-energy production: potential and research challenges. *Renewable and Sustainable Energy Reviews,* 15(9), 4295-4301.

Ashekuzzaman, S.M. & Poulsen, G. (2011). Optimizing feed composition for improved methane yield during anaerobic digestion of cow manure based waste mixtures, *Bioresource Technology*, 102(3), 2213-2218.

Axaopoulos, P. ; Panagakis, P.; Tsavdaris, A. & Georgakakis, D. (2001). Simulation and experimental performance, of a solar-heated anaerobic digester. *Solar Energy*, 70, 155-164.

Axaopoulos, P. & panagakis, P. (2003). Energy and economic analysis of biogas heated livestock buildings. *Biomass and Bioenergy*, 24, 239-248.

Bagi, Z.; Acs, N.; Balint, K.L.; Horvath, L.; Dobo, K.; Perei, K.R.; Rakhely, G. & Kovacs, K.L. (2007). Biotechnological intensification of biogas production. *Appl Microbiol Biotechnol,* 76, 473–482.

Billa, E.; Dimitris, P.; Koullas, B.M. & Koukios, E.G. (1997). Structure and composition of sweet sorghum stalk components. *Industrial Crops and Products*, 6, 297-302.

Börjesson, P. & Tufvesson, L.M. (2011). Agricultural crop-based biofuels – resource efficiency and environmental performance including direct land use changes. *J. Clean. Prod.,* 19, 108–120.

Bruni, E.; Jensen, A.P. & Angelidaki, I. (2010). Comparative study of mechanical, hydrothermal, chemical and enzymatic treatments of digested biofibers to improve biogas production. *Bioresour Technol*, 101, 8713–8717.

Buffiere, P.; Mirquez, L.D.; Steyer, J.P.; Bernet, N. & Delgenes, J.P. (2008). Anaerobic digestion of solid wastes needs research to face an increasing industrial success. *Int J Chem React Eng*, 6,84-94.

Cavinato, C.; Fatone, F.; Bolzonella, D. & Pavan,P. (2010). Thermophilic anaerobic co-digestion of cattle manure with agro-wastes and energy crops: Comparison of pilot and full scale experiences. *Bioresource Technology*, 101, 545-550.

Chandra, R. ; Takeuchi, H. & Hasegawa, T. (2012). Methane production from lignocellulosic agricultural crop wastes: A review in context to second generation of biofuel production. *Renewable and Sustainable Energy Rev*iews, 16, 1462-1476.

Chandra, R. ; Takeuchi, H. & Hasegawa, T. (2012). Hydrothermal pretreatment of rice straw biomass: A potential and promising method for enhanced methane production. *Applied Energy*, 94, 129-140.

De Meester, S. ; Demeyer, J.; Velghe, F.; Peene, A.; Van Langenhove, H. & Dewulf, J.(2012). The environmental sustainability of anaerobic digestion as a biomass valorization technology. *Bioresource Technology*, 121, 396-403.

Demirel, B. and P. Scherer, Trace element requirements of agricultural biogas digesters during biological conversion of renewable biomass to methane, Biomass and Bioenergy, Volume 35, Issue 3, March 2011, Pages 992-998.

Deublein, D. & Steinhauser, A. (2008). *Biogas from Waste and Renewable Resources*; Wiley Online Library: Weinheim, Germany.

EurObserv'ER biogas barometer (2011). http://www.eurobserv-er.org/

FAOSTAT. (2011). Global rice production by year. Food and Agriculture Orga- nization of the United Nations (http://faostat.fao.org).

Fermoso, F.G.; Bartacek, J.; Jansen, S. & Lens, P.N.L. (2009). Metal supplementation to UASB bioreactors: from cell-metal interactions to full-scale application. *Science of the Total Environment,* 407,3652-3667.

Gadde, B.; Menke, C. & Wassmann, R. (2009). Rice straw as a renewable energy source in India, Thailand, and the Philippines: overall potential and limitations for energy contribution and greenhouse gas mitigation. *Biomass Bioenerg,* 33, 1532-1546.

Georgakakis, D.; Christopoulou, N.; Chatziathanassiou, A. & Venetis, T. (2003). Development and use of an economic evaluation model to assess establishment of local centralized rural biogas plants in Greece. *Appl. Biochem. Biotechnol. 109*, 275–284.

Gerardi, M.H. (2003). The microbiology of anaerobic digesters, waste water microbiology series Hoboken, New Jersey: John Wiley & Sons, Inc.

Gonzalez-Gil, G.; Seghezzo, L.; Lettinga, G. & Kleerebezem, R. (2001). Kinetics and mass-transfer phenomena in anaerobic granular sludge. *Biotechnology and Bioen- gineering,* 73, 125-34.

Guiot, S.R.; Frigon, J.C.; Cimpoia, R. & Tartakovsky, B. (2009). Anaerobic digestion : An effective technology for biofuel production. International Workshop on Anaerobic Digestion: an old story for today and tomorrow, December 10-11th, 2009, Narbonne (France).

Hartmann, H. & Ahring, B.K. (2005). Anaerobic digestion of the organic fraction of municipal solid waste: Influence of co-digestion with manure. *Water Research*, 39, 1543-1552.

Hartmann, H. & Ahring, B.K. (2006). Strategies for the anaerobic digestion of the organic fraction of municipal solid waste: an overview. *Water Science and Technology*, 53, 7-22.

Hendriks, A. & Zeeman, G. (2009). Pretreatments to enhance the digestibility of lignocellulosic biomass. *Bioresour Technol,* 100, 10–8.

Hill, D.T. (1983). Simplified monod kinetics of methane fermentation of animal wastes. *Agricultural Wastes,* 5, 1-16.

Holm-Nielsen, J.B.; Al Seadi, T. & Oleskowicz-Popiel, P. (2009). The future of anaerobic digestion and biogas utilization. *Bioresour. Technol.* 1000, 5478–5484.

Hossain, A.K. & Badr, O. (2007). Prospects of renewable energy utilization for electricity generation in Bangladesh. *Renew Sust Energ Rev* , 11, 1617-1649.

Hublin, A.; Zokic, T.I. & Zelic, B. (2012). Optimization of Biogas Production from Co-digestion of Whey and Cow Manure. Biotechnol. Bioprocess Eng., 17, 1284-1293.

Jiang, Y.; Heaven, S. & Banks, C.J. (2012). Strategies for stable anaerobic digestion of vegetable waste. *Renewable Energy*, 44, 206-214.

Lee, J.M. (1992). Biochemical Engineering, Prentice-Hall Inc., 1992, ISBN 0-13-085317-8.

Lin, J; Zuo, JE; Gan, LL; Li, P; Liu, FL; Wang, KJ; Chen, L; Gan, HN (2011). Effects of mixture ratio on anaerobic co-digestion with fruit and vegetable waste and food waste of China. Journal of Environmental Science-China, 23, 1403-1408.

Ma, J.; Yu, L.; Frear, C.; Zhao, Q. ; Li, X. & Chen, S. (2013). Kinetics of psychrophilic anaerobic sequencing batch reactor treating flushed dairy manure. *Bioresource Technology*, 131, 6-12.

Madsen, M. ; Holm-Nielsen, J.B. & Esbensen, K. (2011). Monitoring of anaerobic digestion processes: A review perspective. *Renewable & Sustainable Energy Reviews,* 15, 3141-3155.

Macias-Corral, M.; Samani, Z.; Hanson, A.; Smith, G.; Funk, P.; Yu, H. & Longworth, J. (2008). Anaerobic digestion of municipal solid waste and agricultural waste and the effect of co-digestion with dairy cow manure. *Bioresource Technology*, 99, 8288-8293,

Mata-Alvarez, J.; Llabrés, P.; Cecchi, F. & Pavan, P. (1992). Anaerobic digestion of the Barcelona central food market organic wastes: experimental study. *Bioresource Technology* , 39, 39-48.

Mata-Alvarez, J.; Macé, S. & Llabrés, P. (2000). Anaerobic digestion of organic solid wastes. An overview of research achievements and perspectives. *Bioresource Technology,* 74, 3-16.

Mata-Alvarez, J.; Dosta, J.; Macé, S. & Astals, S. (2011). Codigestion of solid wastes: A review of its uses and perspectives including modeling. *Critical Reviews in Biotechnology,* 31, 99-111.

McGraw-Hill Dictionary of Scientific & Technical Terms, 6E, Copyright © 2003 by The McGraw-Hill Companies, Inc.

Menardo, S. & Balsari, P. (2012). An Analysis of the Energy Potential of Anaerobic Digestion of Agricultural By-Products and Organic Waste. *Bioenergy Research*, 5, 759-767.

Mendes, A.A.; Pereira, E.B. & Castro, H.F. (2006). Effect of the enzymatic hydrolysis pretreatment of lipids-rich wastewater on the anaerobic biodigestion. Biochem Engin J, 32, 185–190.

Merlin G.; Kohler, F.; Bouvier, M.; Lissolo, T. & Boileau, H. (2012). Importance of heat transfer in an anaerobic digestion plant in a continental climate context. *Bioresource Technology*, *124*, *59-67.*

Misi, S.N. & Froster, C.F., (2001). Batch co-digestion of multi-component agro-wastes. *Bioresour. Technol.,* 80, 19–28.

Moletta, R. (2008). La méthanisation, Tec-Doc Lavoisier, ISBN978-2-7430-1036-2

Moletta, R. (2009). Le traitement des déchets, Tec-Doc Lavoisier, ISBN 978-2-7430-1108-6.

Moller, H.B.; Sommer, S.G. & Ahring, B.K. (2004). Methane productivity of manure, strawand solid fractions of manure. *Biomass Bioenergy,* 26, 485–95.

Moser, A. (1988). Bioprocess technology, kinetics and reactors, Springer Verlag, ISBN 0-387-96603-X.

Mussoline, W. ; Esposito, G.; Giordano, A. & Lens, P. (2013). The Anaerobic Digestion of Rice Straw: A Review. *Critical Reviews in Environmental Science and Technology,* 43, 895-915.

Nallathambi, G.V. (2004). Biochemical methane potential of fruits and vegetable solid waste feedstocks. *Biomass Bioenergy*, 26, 389–99.

Nielsen, H.B. & Angelidaki, I. (2008). Codigestion of manure and industrial organic waste at centralized biogas plants: process imbalances and limitations, *Water Sci. Technol.*, 58, 1521-1528.

Osuna, M.B.; Zandvoort, M.H.; Iza, J.M.; Lettinga, G. & Lens, P.N.L.(2003). Effects of trace element addition on volatile fatty acid conversions in anaerobic granular sludge reactors. *Environmental Technology* , 24, 573-87.

Parawira, W. (2012). Enzyme research and applications in biotechnological intensification of biogas production. *Critical Review in Biotechnology*, 32, 172-186.

Perez, J.; Dorado, J.M.; Rubia, T.D. & Martinez, J. (2002). Biodegradation and biological treatment of cellulose, hemicellulose and lignin: an overview. *Journal of International Microbiology,* 5, 53–63.

Petersson, A.; Thomsen, M.H.; Hauggaard-Nielsen, H. & Thomsen, A.B. (2007). Potential bioethanol and biogas production using lignocellulosic biomass from winter rye, oilseed rape and faba bean. *Biomass Bioenergy* , 31, 812–819.

Rahman, M.; Mizanur, J.V. & Paatero, A. (2012). Methodological approach for assessing potential of sustainable agricultural residues for electricity generation: South Asian perspective. *Biomass and Bioenergy*, 47, 153-163.

Rajendran, K.; Aslenzadeh, S. & Taherzadeh, M. (2012). Household biogas digester, a review. *Energies*, 5, 2911-2942.

Ravindranath, N.H.; Somashekar, H.I.; Nagaraja, M.S.; Sudha, P.; Sangeetha, G. & Bhattacharya, S.C. (2005). Assessment of sustainable non-plantation biomass resources potential for energy in India. *Biomass Bioenerg.*, 29, 178-190.

Raynal, J.; Delgenès, J.P. & Moletta, R. (1998). Two-phase anaerobic digestion of solid wastes by a multiple liquefaction reactors process. *Bioresource Technology*, 65, 97-103.

Richarson, J.F. & Peacock, D.G. (1994). Chemical Engineering, Vol 3, third edition, Chemical & biochemical Reactors & Process Control, ISBN 0-08-041002-2.

Sheng, K.C.; Xia, Y. H.; Gharibi, A. & Chen, X. (2012). Examining a Hybrid plug-flow Pilot reactor for Anaerobic Digestion of Farm-Based Biodegradable Solids International Journal of Environmental Research, 6, 335-344.

Sinbuathong, N.; Sirirote, P.; Sillapacharoenkul, B.; Munakata-Marr, J. & Chulalaksananukul, S. (2012). Biogas Production from Two-stage Anaerobic Digestion of Jatropha curcas Seed Cake. *Energy Sources Part A-Recovary Utilization and Environmental Effects*, 34, 2048-2056.

Symons, G.E. &, Buswell, A.M. (1993). The methane fermentation of carbohydrates. *J. Am. Chem. Soc.*, 55, 2028-36.

Taherzadeh, M.J. & Karimi, K. (2008). Pretreatment of lignocellulosic wastes to improve ethanol and biogas production: a review. *International Journal of Molecular Science,* 9, 1621–1651.

Valladao, A.B.G.; Freire, D.M.G. & Cammarota, M.C. (2007). Enzymatic pre- hydrolysis applied to the anaerobic treatment of effluents from poultry slaughterhouses. *Int Biodeterior Biodegrad,* 60, 219–225.

Van Lier, J.B.; Mahmoud, N. & Zeeman, G. (2008). Anaerobic wastewater treatment. In biological Wastewater Treatment, Principles, Modeling and design. IWA Publishing, London, UK, 415-456.

Vervaeren, H.; Hostyn, K.; Ghekiere, G. & Willems B. (2010). Biological ensilage additives as pretreatment for maize to increase the biogas production. *Renew Energ*, 35, 2089–2093.

Wang, S.; Chandrasekhara Rao, N.; Qiu, R. & Moletta, R. (2009). Performance and kinetic evaluation of anaerobic moving bed biofilm reactor for treating milk permeate from dairy industry. *Bioresource Technology*, 100, 5641-5647.

Ward, A.J.; Hobbs, P.J.; Holliman, P.J. & Jones, D.L. (1999). Review:optimisation of the anaerobic digestion of agricultural resources. *Bioresour Technol.,* 79, 28–40.

Weiland, P. (2006). Biomass digestion in agriculture: a successful pathway for the energy production and waste treatment in Germany. *Eng. Life Sci.*, 6, 302–309.

Wilkie, A.C. (2005). Anaerobic Digestion of Dairy Manure: Design and Process Considerations. In Dairy Manure Management: Treatment, Handling, and Community Relations NRAES-176 p.301-312. Natural Resource, Agriculture, and Engineering Service Cornell University, Ithaca, NY, 2005.

Wint, W.; & Robinson, T. (2007). Gridded Livestock of the World. FAO, Rome.

Worm, P.; Fermoso, F.G.; Lens, P.N.L. & Plugge, C.M. (2009). Decreased activity of a propionate degrading community in a UASB reactor fed with synthetic medium without molybdenum, tungsten and selenium. *Enzyme and Microbial Technology*, 45, 139-145.

Yettou, A. ; Miri, R. & Belhamel, M. (2007). Étude expérimentale d'un digesteur de production de biogaz alimenté en énergie par un chauffe-eau solaire. *Revue des Énergies Renouvelables* , CER'07, Oujda, 13-17

Zhang, Z.; Gonzalez, A. M.; Davies, E.G. R. & Liu, Y. (2012). Utilization of byproducts and waste materials from meat, poultry and fish processing industries: a review. *Water Environment Research,*84, 1386-1406.

In: Anaerobic Digestion
Editor: Ana Torales

ISBN: 978-1-62808-884-7

Chapter 3

EVALUATION OF DECENTRALIZED ANAEROBIC DIGESTION SYSTEMS FOR WASTEWATER TREATMENT, REUSE AND NUTRIENT RECYCLING TO ALLEVIATE WATER AND FOOD STRESS

***M. Duqqah*[1,2,*] *and U. Zaher*[2,†]**
[1]Department of Land, Water and Environment,
University of Jordan, Amman, Jordan
[2]Department of Biological Systems Engineering,
Washington State University, Pullman, WA, US

ABSTRACT

Food shortage and drought is a growing concern in developing countries requiring economic reuse and nutrients recycling. Anaerobic digestion can be economically applied to treat domestic wastewater from rural and small communities. Upflow Anaerobic Sludge Blanket (UASB) is proven low-cost technology that can be sustainably built at small scale and locally operated for decentralized treatment and reuse. This book chapter evaluates typical decentralized UASB implementation in integration with crop production. The anaerobic process was evaluated for nutrient recovery, effluent reuse in drip irrigation system and for biogas production. N, P and K release in the effluent waters were evaluated as alternative fertilizer for tomato production in green house using drip irrigation. The UASB was followed by Rotated Biological Contactors (RBC), solids separation and filtration post treatment units to control chemical oxygen demand (COD) and suspended solids (SS) in the irrigation system. Biogas production was evaluated as a renewable source for energy generation and heat supply on a household scale. The health and environmental impacts were evaluated in the irrigation experiment according to the World Health Organization (WHO) and the Food Agriculture organization (FAO) guidelines.

[*] M. Duqqah: Department of Land, Water and Environment, The University of Jordan, Amman 11942, Jordan. E-mail: mmduqqah@ju.edu.jo.

[†] U. Zaher: Department of Biological Systems Engineering, Washington State University, P.O. Box 646120, Pullman, WA 99164-6120, US. E-mail: mm.duqqah@wsu.edu, zaheru@wsu.edu.

The characteristics of the UASB effluent after post-treatment were: EC 3.0 dS/m, pH 7.7, BOD 303 mg/l, NO_3 5.1 mg/l, NH_4 54.7 mg/l, P 18.0 mg/l, K 36.8 mg/l, E. coli 31.7E+5 MPN/100ml and T. coliform 68.7E+5 MPN/100ml. These characteristics sufficed plant nutrient needs and resulted in 12% higher tomato crop yields compared to fertigation with inorganic fertilizer. However, salinity level was severe according to FAO. Although effluent counts of pathogenic indicators were not acceptable according to WHO, the counts were significantly reduced after irrigation as indicated by the analytical results of Escherichia coli and Total coliform in soil (<1.2E+3) and in plant (<1 MPN/100 ml).

Keywords: UASB, decentralized, reuse, nutrient cycling, pathogens

INTRODUCTION

Water for irrigation and food production constitutes one of the greatest pressures on freshwater resources. Agriculture accounts for about 70% of global freshwater withdrawals (up to 90% in some fast-growing economies) (WWAP, 2012). The world population is predicted to grow from 6.9 billion in 2010 to 9.1 billion in 2050 (UNDESA, 2009a). With expected increases in population, by 2050, food demand is predicted to increase by 70% (Bruinsma, 2009), while energy demand from hydropower and other renewable energy resources will rise by 60% (WWAP, 2009). Concerns about food insecurity are growing across the globe and more water will be needed to meet increasing demands for food and energy (biofuels). Globally, irrigated crop yields are about 2.7 times those of rainfed farming, hence irrigation will continue to play an important role in food production.

Society no longer has the luxury of using water only once. In fact, in many parts of the world, using water only once is no longer an option (Levine and Asano, 2004). As we chart a path in the 21st century, we contend that wastewater contains resources worthy of recovering and that the development of technologies, practices, and policies that enable cost-effective recovery will have broad geopolitical implications (Guest et al., 2009). Efficient and environmentally sustainable reuse of wastewaters must be implemented whenever possible. Sustainability demands that we acknowledge wastewater as a renewable resource from which water, nutrients, and energy can be recovered for reuse (Verbyla et al., 2013).

In order to help address these issues, wastewater is now being looked at more as a resource than as a waste, a resource for water, for energy, and for the plant fertilizing nutrients, nitrogen (N), phosphorus (P) and potassium (K) (Asano et al., 2007). Use of wastewater for landscape and crop irrigation is a widely accepted and growing practice to save water and to make use of the fertilizing elements it contains. Similarly, use of wastewater as a source of energy has a long history, especially through the anaerobic conversion of wastewater' s organic content into methane (CH_4) gas, a useful biofuel (Lettinga et al., 1983, Lettinga, G, 2010). According to the United States Environmental Protection Agency (USEPA, 2011), wastewaters accounted for more than 6%, or approximately 450 million metric tones of CO_2 equivalent of the estimated global anthropogenic methane emissions in 2010. These, together with manure management, represent 35% of methane emissions generated by all the wastes (liquid, solid, and gaseous). Given the enormous, and rising, quantities of methane emissions that are involved it is imperative that steps are taken to capture that methane.

Besides controlling global warming, generating a relatively cleaner source of energy, and improving public hygiene, these steps would also enable capturing plant nutrients (principally N, P, and K but also a host of micronutrients) which otherwise are not only wasted but become a major source of water pollution (Tauseef et al., 2013). Anaerobic digesters promise to facilitate such initiatives.

By shifting away from today's paradigm, which focuses on what must be removed from wastewater, to a new paradigm focusing on what can be recovered, sanitation systems may begin to be described as resource recovery systems, a conceptual transformation that could allow the perceived impact of wastewater on communities to become a net positive (Guest et al., 2009). Furthermore, it is important to evaluate resources recovery in a way that links sanitation goals with water management and food production. To foster sustainability, the quality and quantity of water use must be linked to water availability. Cost-effective use of wastewater for irrigation applications necessitates producing it relatively close to the potential user. In addition, because requirements for irrigation tend to be seasonal whereas wastewater production is continuous, effective storage also is needed. One concept that is gaining acceptance is decentralized wastewater treatment facilities, which produce wastewater in locations that might be more readily accessible for use by agriculture (Crites et al., 1998). This approach avoids the cost of transporting water to a central facility and then returning the wastewater to another location.

Centralized and decentralized wastewater treatment systems have coexisted over the past years. Centralized wastewater treatment is often unable to make use of the treated wastewater and instead of being recycled it is discharged into rivers. Furthermore, when inadequately treated wastewater is disposed to the soil or reused for irrigation, it can lead to the development of hydrophobicity in receiving soils and subsequent poor distribution of irrigation water (Wallach et al., 2005; Tarchitzky et al., 2007; Travis et al., 2010); fecal bacteria health risks (Jamieson et al., 2002; Travis et al., 2010); and particularly in arid regions, salinization of soils (Rebhun, 2004). In contrast, the wastewater that is generated in rural areas or in small communities served by decentralized systems may be used for groundwater recharge, and polished by physical, chemical and biological processes during soil passage. Alternatively, the treated wastewater may be sent to a nearby surface water body, or used for irrigation. The sludge is converted into compost that can be used, on site, as a soil conditioner and as a fertilizer source in gardening or landscaping. Decentralized wastewater treatment systems must meet health and environmental requirements for disposal or reuse, and typically must provide advanced wastewater treatment highly effective, robust, easy to operate, and low in costs, and easy to maintain and operated in order to be successfully applied in rural areas.

Upflow Anaerobic Sludge Blanket (UASB) is proven low-cost technology that can be sustainably built at small scale and locally operated for decentralized treatment and reuse (Lettinga et al., 1980, 1993; Tawfik et al., 2003; Elmitwalli et al., 2003; Tandukar et al., 2005). They have been tested for domestic, agricultural, and industrial wastewaters, (Oktem et al., 2008; Yetilmezsoy et al., 2008; Mahmoud, 2008). The aim of this research was to develop and test a combined anaerobic and Rotated Biological Contactors (RBC) on-site treatment of domestic wastewater and to study the effect of irrigation by UASB effluent, as compared to normal water, on the yield and quality and pathogenic contamination of tomato fruit grown under greenhouse by using pressurized drip irrigation system in an arid environment.

MATERIALS AND METHODS

1. Experimental Site

The experimental site is located close to Abu Nuiesser wastewater treatment plant near the capital city Amman, Jordan. The climate is semi-arid as characterized by Mediterranean bioclimatic zoning. The rainfall ranges from 400-600 mm with minimum annual temperature ranges from 5-10 °C and maximum annual temperature ranges from 25-35 °C. The rainy season extends from October to mid of May. The annual potential evaporation varies from 1900-2200 mm. About 70% of the annual evaporation is recorded in the dry season from May to October. The soil texture was identified using the relative percentage proportions of sand, silt and clay by using the United States Department of Agriculture (USDA) soil texture classification (Soil Survey Staff, 2011).

2. Experimental Design

The experiment was conducted in a $500m^2$ greenhouse built 60 m downstream of the Abu-Nusier treatment plant. The land was ploughed twice before installing the greenhouse. The land was divided into 2.5m by 2.5m plots with paths of 0.5 m between blocks. The plots were randomly distributed on the basis of three treatments with three replicates for each irrigation water source. This paper compares three irrigation water sources: (1) Effluent of Upflow Anaerobic Sludge Blanket followed by Rotated Biological Contactors (UASB-RBC), (2) Fresh water with fertilizer (FWF), and (3) Fresh water as a control (FW).

Every plot contained 2 laterals from the main of a drip irrigation system. Every lateral contains 5 drippers supplying 20 l/hr. The soil was initially irrigated with fresh water (source 3) until saturation and the laterals were covered with plastic mulch to minimize the contact between irrigation water and crop and labor, to minimize water evaporation losses from the soil and to control weed germination.

A pressurized drip irrigation system consisted of storage tanks for each water quality followed by a head unit that consist of sand filter, fertilizer injection unit (for source 2), disc filter and valves to control water follow to the green house. At the inlet to greenhouse pressure regulator valves were installed to control the pressure required for drippers. Main line (32 mm diameter) will carry water to a set of laterals (20 mm diameter) installed in the field to irrigate tomato plants.

A UASB reactor of 20 l effective volume was used in the experiment. Five sampling ports were placed along the height of the reactor. Temperature was under control through circulating thermostat water in a water jacket that surrounds the reactor. The reactor was inoculated with 10L of flocculent anaerobic sludge, originating from a pilot UASB reactor operating in Abu-Nussier wastewater treatment plant serving community's residence hall. Influent preserved at 4C, was supplied to the reactor by peristaltic pump using timer to fix the flow rate. The reactor operation was launched under operation loading rate of 3.22 kg COD/ m^3.d, hydraulic retention time (HRT) of 12 hr and influent concentration of 1.6 g COD/l.

The experiment took place during dry season and carried out on tomato crop. One-month-old tomato seedlings were replanted from nursery to experiment plots. All plots were treated against insects, fungi and nematodes through the life cycle of the plant.

3. Irrigation and Fertigation

The amount and frequency of irrigation were the same for every treatment. Irrigation amounts were computed according to pan evaporation measurement, which installed inside greenhouse for the whole experiment period. The irrigation was carried out two times a week for two hours for the whole growing period of tomato plant.

For fertilized treatment (FWF) mineral fertilizers were applied according to a survey of the farming practices in Jordan. Total fertilizers added per hectare were: 59.4 Kg N/ha in the form of nitrate (NO_3), 27.7 Kg P/ha in the form of P_2O_5 minerals and 115.4 Kg K/ha in the form of K_2O. NPK fertilizer was added as 20:20:20 and 12:12:36 according to the plant needs during each growth stage after transplanting tomato plants.

4. Sampling and Analysis

Water samples from FW and UASB-RBC effluent were analyzed for BOD_5, mineral nitrogen, PO_4^{-2}, K^+, Na, pH, EC, and trace elements. Water samples were taken at the beginning of the experiment and prior to each irrigating from a storage tank after stirring to analyze water quality and record observations with regard to color, odor, and algae. Water samples were collected in tight cap closure bottles, which were clean and free of oil, slats and other contaminants. Each bottle is rinsed with the sampled water and filled completely to rim. Samples were refrigerated at 4°C and analyzed as soon as possible for the above parameters according to procedures of the Standard Method for the Examination of Water and Wastewater (APHA, 1998). NH_4 and NO_3 were measured using spectrophotometer at wavelength 206 nm. Phosphorus was analyzed as orthophosphate using spectrophotometer at wavelength 700nm. Potassium K was measured using standard flame photometer. *Total coliforms (T. coliform), Escherichia coli (E. coli)* and *Enterococcus* were also measured in fresh water and treated wastewater samples. Samples were collected for these biological quality parameters in sterile bottles, preserved with sodium thiosulfate to neutralize residual chlorine and transferred directly to the lab for analysis according to Standard Method for the Examination of Water and Wastewater (APHA, 1998).

Soil samples were collected from the experiment area before planting to test homogeneity of the irrigated area and soil physicochemical properties. Initial nutrients were analyzed in the collected samples. Also, undisturbed soil column samples were collected from the greenhouse test plots at the middle and the end of the experiment, 2 days after irrigation. For each plot irrigated with designated water sources composite sample samples were collected at depths: 0-15 cm, 15-30 cm, 30-45 cm and 45-60 cm. All soil samples were allowed to dry for a day then were ground in a mechanical mortar and pestle and passed through a number 12-sieve (approximately 2 mm screen). EC, pH, total nitrogen, ammonium (NH_4), nitrate (NO_3), extractable phosphorus, and extractable potassium were measured.

NH_4 and NO_3 were determined using continuous flow analysis (CFA) system by extracting NH_4 and NO_3 from the soil by KCl (0.1 M) and then measured by spectrophotometer at wavelength 660 and 540 nm respectively (Ryan et al., 1996). Total nitrogen was measured by Kjeldahl procedure (Ryan et al., 1996).

Available phosphorous was extracted using sodium bicarbonate solution since Jordan soil's is mainly calcareous. Phosphorous concentration was then measured using a standard spectrophotometer and a wavelength of 690 nm (Ryan et al., 1996). Potassium was extracted in 1.0 N ammonium acetate solution and then potassium was quantified in the extracted solution using a flame photometer (Ryan et al., 1996).

For biological parameters, samples were analyzed after both 5 and 10 days from the last irrigation event. Biological parameters of *T. coliforms, E. coli* and *Enterococcus* were quantified in samples from each depth based on the Standard Methods (APHA, 1998).

At the end of growing season tomato plant (stems, leaves and fruits) from bottom, middle and top of the plant were collected, weighed and analyzed for total N, P and K. The specific parts of the plant were removed by using clasp knife and stored in zip bags during transportation from field to laboratory.

In the laboratory the samples were washed with distilled water in order to remove dust, pesticides and other residues. Then the samples were air dried first then dried in an oven at 65°C for 24 hours to stop enzymatic activity. The dry samples were mechanically grounded in a stainless steel mill. Nutrients were analyzed after digestion of plant material in a sulfuric-salicylic acid mixture followed by distillation.

Total N was measured by titration according to Ryan et al. (1996). Was measured calorimetrically using spectrophotometer and wavelength of 410 nm (Ryan et al., 1996). The total K in the plant samples was analyzed by dry ashing and quantification using flame photometer (Ryan et al., 1996). *T. coliforms, E. coli, Enterococcus* and *Salmonella* were also analyzed for leaves and fruits based on the Standard Method (APHA, 1998).

5. Mass Balance

Efficiency of nutrient uptake and sequestration was evaluated by mass balance of N, P and K in the top two layers of the 2.5m x 2.5m test plots for the three irrigation and fertigation sources. The mass balance was evaluated according to equation (1):

$$[mass\ balance\ (N_i)]_{source} = \sum_{layer=1}^{2} soil\ intial\ (N_i) + \left[added(N_i) - \sum_{part=1}^{3} Plant\ uptake(N_i) - \sum_{layer=1}^{2} soil\ final\ (N_i) \right]_{source} \quad (1)$$

where: source = {FW, FWF, AS}; i={N, P, K} ; layer ={layer 0-15cm, layer 15-30cm} ; part ={stem, leaves, fruit}

RESULTS

1. Quality of Irrigation Water

Table 1 shows the irrigation water characteristics for both FW and UASB-RBC effluent sources in comparison with FAO and WHO recommendation limits and guidelines. Generally, both sources were suitable for irrigation with slight to moderate restriction according to FAO guidelines, while the biological characteristics of pathogenic indicators of UASB-RBC effluent were not suitable for irrigation according to WHO guidelines for wastewater use in agriculture.

According to WHO standards, wastewater with the listed qualities should not be used even for restricted irrigation, e.g. to irrigate fodders. The WHO reported that the maximum limit of *fecal coliform* in wastewater used for restricted irrigation should be 1000 MPN/ 100ml.

Also, additional biological quality measures show high numbers of *Enterococcus* but no detection of *Salmonella* in UASB-RBC effluent. EC, Na and Cl measurements were higher indicating higher salinity of the UASB-RBC effluent but within the FAO limits.

Trace metals were also within the FAO limits for UASB-RBC effluent but not detected in the FW indicating that UASB-RBC effluent contained appropriate amount of micronutrients. UASB-RBC effluent macronutrient content of N, P, K was higher in UASB-RBC effluent with moderate nitrate concentrations compared to FAO guidelines. There was no guideline for ammonium concentration but UASB-RBC effluent was relatively higher in ammonium content compared to FW.

2. Initial Soil Analysis

The soil is classified as silty-clay-loam according to the soil texture classification of United States Department of Agriculture (USDA). Table 2 shows the initial soil analysis for all sampled depths. The EC_e, Ca, Mg and Cl analysis indicate low soil salinity along the soil profile.

The soil has high alkalinity with higher cation concentration in the top 30 cm and pH above 8 for the top 60 cm soil profile. N and P were naturally adequate for crop growth with lower concentrations in deeper soil layers. Similarly, C/N ratio was 50% higher in the top 30 cm compared to deeper layers. K was extremely high in soil with decreasing trend with depth.

The fresh and dry weights of fruits, stems, leaves and whole plants for each treatment are shown in Figure 1. Reuse of the UASB-RBC effluent in tomato plant irrigation doubled the fruit yield as well as the total dry weight compared to other irrigation water sources, FW and FWF.

FWF results indicate that adding fertilizers to FW according to the farmer practices increased the tomato fruit yield slightly. The influence of fertilizer addition on vegetative growth was minimal compared to FW only as indicated by dry weight results. Generally the dry weight yields were in establishing vegetative growth.

Table 1. Chemical and biological analysis of irrigation water

Parameter	Units	FW	UASB-RBC	Guidelines and Recommendation
FAO Guideline for interpretation of water quality for irrigation:[*]				
pH		7.86	7.78	Normal range (6.5-8.0)
EC	dS/m	1.03	2.10	Slight to moderate (0.7-3.0)
Na	me/l	2.77	9.00	Not specified for drip irrigation, but for sprinkler and surface it considered as severe for UASB-RBC (>9.0)
Cl	me/l	4.00	10.00	Not specified for drip irrigation, but for sprinkler and surface it considered as severe for UASB-RBC (>10.0)
B		-----	-----	None (<0.7)
NO_3-N	mg/l	2.60	5.12	None for FW (<5.0) and Slight to moderate for UASB-RBC (5.0-30.0)
FAO Recommendation maximum concentration of trace elements in irrigation water:[*]				
Fe	mg/l	-----	1.67	Below maximum concentration (5.0)
Zn	mg/l	-----	0.51	Below maximum concentration (2.0)
Ni	mg/l	-----	0.03	Below maximum concentration (0.2)
Mn	mg/l	-----	0.08	Below maximum concentration (0.2)
Cd	mg/l	-----	<0.01	Below maximum concentration (0.01)
Cu	mg/l	-----	0.04	Below maximum concentration (0.2)
WHO Recommended minimum verification monitoring of microbial performance targets for wastewater and excreta use in agriculture:[**]				
E. coli	MPN/100ml	-----	3.20E+06	Not Acceptable for unrestricted irrigation with drip irrigation and high growing crops (≤1.0E+05) and for restricted irrigation with labor intensive and high contact agriculture (≤1.0E+04)
T. coliform	MPN/100ml	-----	8.90E+06	Not Acceptable (≤1000)
Additional quality parameters				
NH_4^+-N	mg/l	2.65	54.68	
P	mg/l	0.11	18.11	
K	mg/l	5.75	36.76	
Enterococcus	MPN/100ml	-----	7.60E+02	

[*]Adopted from Ayers and Wescot, FAO irrigation and drainage paper 29 Rev. 1, 1994.

[**]Adopted from WHO guidelines for the safe use of wastewater, excreta and grey-water: volume 1, policy and regulatory aspects, 2006.

Table 2. Initial soil analysis

Parameter	Units	Soil depth (cm)			
		0-15	15-30	30-45	45-60
EC_e	dS/m	0.43	0.39	0.34	0.32
pH		8	8	8.2	8.2
C/N		7.7	7.2	5.8	5.1
Ca (Extractable)	ppm	192.6	101.7	67.1	106.1
Mg (Extractable)	ppm	17.7	10.3	7.1	10
Alkalinity as $CaCO_3$	%	27.6	25.1	29.8	30
NO_3-N	mg/kg	29	26	5	15
NH_4^+-N	mg/kg	20	17	17	15
TN	mg/kg	49	42	22	29
P (Extractable)	mg/kg	33	19	16	8
K (Extractable)	mg/kg	700	565	442	360
Cl	ppm	85.2	49.7	21.3	53.3

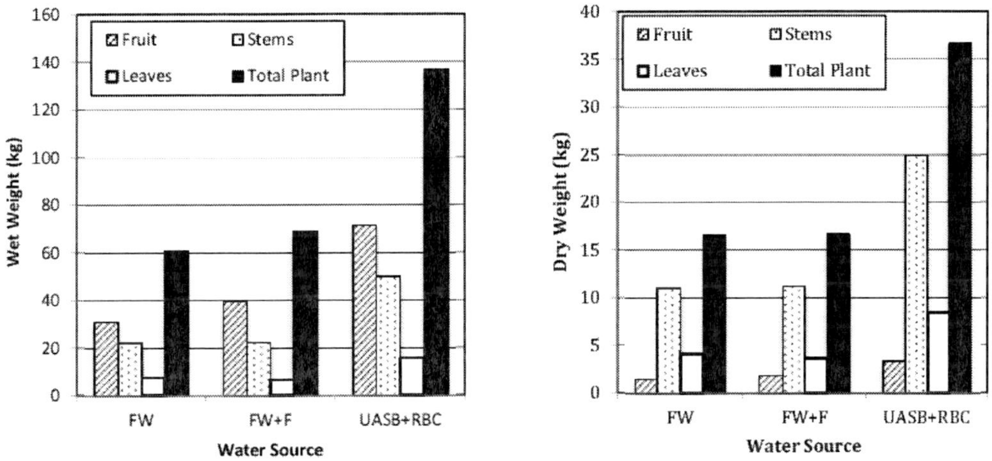

Figure 1. Wet yield (left) and dry yield (right) of plant parts per 2.5m×2.5m test plots comparing: Fresh water (FW), Fresh water and fertilizer (FW+F) and anaerobic effluent from UASB+RBC.

3. Biogas and Crop Yields

The biogas production from the pilot UASB was 30 l/day (1.5 m^3/m^3 reactor/ day) at 80 % methane. This is equivalent to 0.6 m^3 CH_4/m^3 reused wastewater.

4. Nutrients Fate

The fate of macronutrients (N, P and K) in plant and soil as well as the mass balances of the three elements will be presented in the following subsections.

The main result was that reuse of the UASB-RBC effluent maintained the initial nutrient levels in the soil while supplying adequate plant needs. Major results of each pool will be described in the following subsections.

4.1. Nutrients in Plant

The amounts of nutrients (N, P and K) uptake in stems, leaves and fruits of tomato plant for all treatments are presented in Table 3.

The results showed that the N uptake by the leaves was higher than for the stem and fruits that is completely different from the uptake of P by tomato plant where the uptake of P was higher in the fruits and stems. In case of K uptake it was higher in fruits than K uptake by stem and leaves, which shows equal values. The reuse of UASB-RBC effluent and adding fertilizer to FW increase the uptake of N by the leaves and fruits of tomato plant, while the uptake of N by the stem was the same for all treatments. This indicates that N is a critical element during vegetative growth and fruit production for tomato plant. In case of P uptake, the results show that the reuse of UASB-RBC effluent increases the absorption ratio of P by fruits, stems and leaves. The K uptake by tomato plants shows no differences in the amounts of absorbed K in all treatment indicating sufficient amounts of K in soil that are naturally available for plant uptake.

4.2. Nutrients in Soil

Soil samples were analyzed for nitrogen and extractable potassium and phosphorus for different treatments and at different depths (Figure 2). Generally, the uptake of N, P, and K after 30 cm depth where the roots of tomato plant are well developed, was high and causing a sharp decrease in the concentration of N, P, and K in the soil. Reuse of UASB-RBC treated effluent maintained the N, P and K contents that were initially available in the experimented virgin soil.

Furthermore, N bottom figure, was increases in the two middle layers around the root zone, i.e. depth 15-45 cm compared with initial soil N. Irrigating with fresh water with and without fertilizer additions (FW and FWF) had the same effect of degrading the initial soil N, P and K contents. For P concentration, top left figure, was decreasing along the soil profile when using FW and FWF, while the reuse of UASB-RBC effluent shows an increase in the concentration of P in the top 0-15 cm layer. Top right figure shows a decrease in the concentration of K through the soil profile when using FW and FWF but reuse of UASB-RBC effluent maintains the level of K in the soil to its initial concentration.

Table 3. Nutrient fate in fruit, leave and stem for each water source treatment: anaerobic effluent (UASB-RBC), inorganic fertilizer (FWF) and fresh water (FW)

Irrigation water source	Fruit			Leave			Stem		
	N	P	K	N	P	K	N	P	K
Anaerobic effluent (UASB+RBC)	5.3	11.4	55.7	16.1	2.7	24.7	8.0	9.3	27.6
Inorganic Fertilizer (FW+F)	7.4	10.0	59.0	14.5	2.2	24.4	7.8	8.3	31.5
Fresh Water (FW)	3.8	10.2	57.7	9.1	2.5	27.2	8.0	8.5	29.2

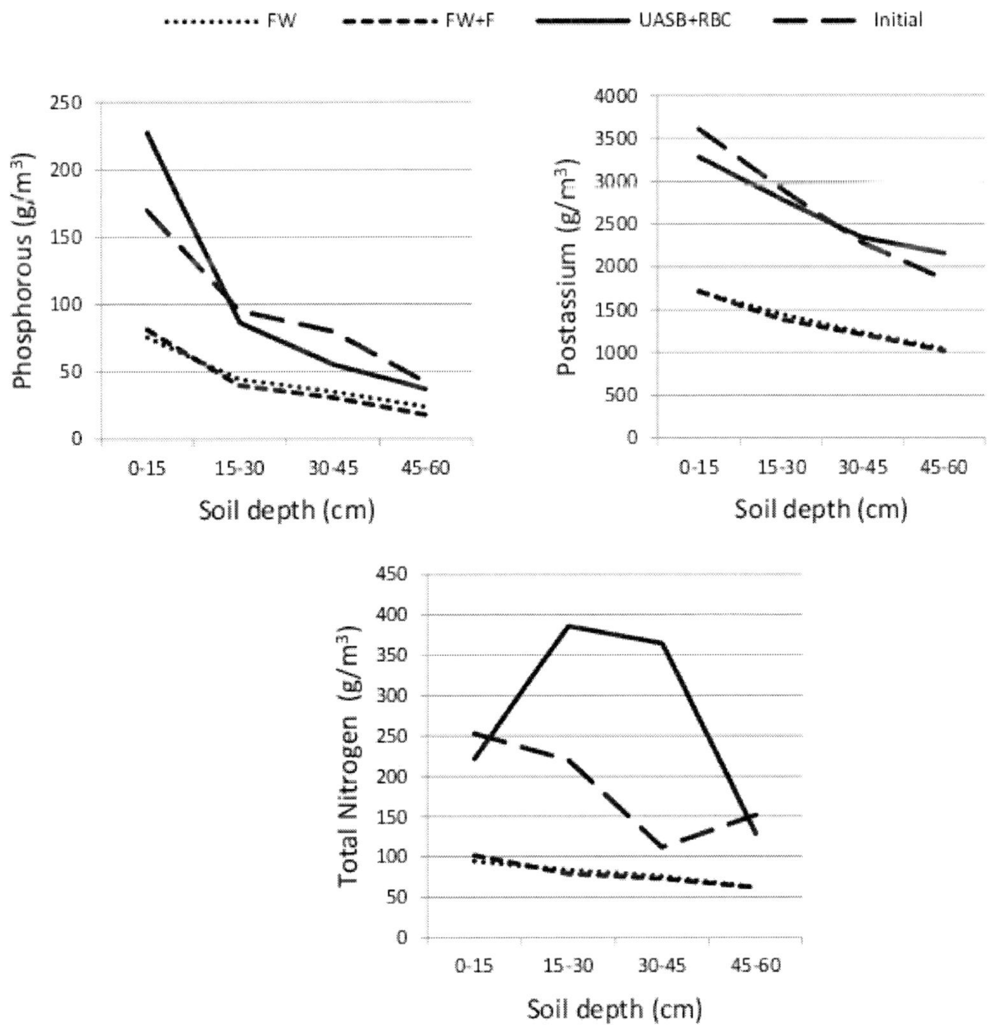

Figure 2. Phosphorous (top left), potassium (top right) and total nitrogen (bottom) fate in soil profile for each resource treatment: anaerobic effluent (UASB-RBC), inorganic fertilizer (FWF) and fresh water (FW).

Figure 3 shows the accumulation of NH_4^+ and NO_3^- through the analyzed soil layers. The reuse of UASB-RBC effluent increases NH_4^+ and NO_3^- accumulations in the soil profile above its initial level in the soil. The initial soil NO_3^- concentrations showing decreasing trend followed by increase in the lower layer 45-60 indicates that NO_3^- is naturally leaching. NO_3^- leaching is caused by irrigation water or rainfall to lower soil layers.

4.3. Nutrients Mass Balance

The mass balance of nutrients N, P and K in the first two layers is presented in Figure 4 to indicate the nutrient utilization around the root zone. Generally, the negative values of the balance indicates that the inputs from each irrigation source and what was initially available in the top 30 cm soil were slightly less than the sum of what was utilized by the plants and what remained in that top layer.

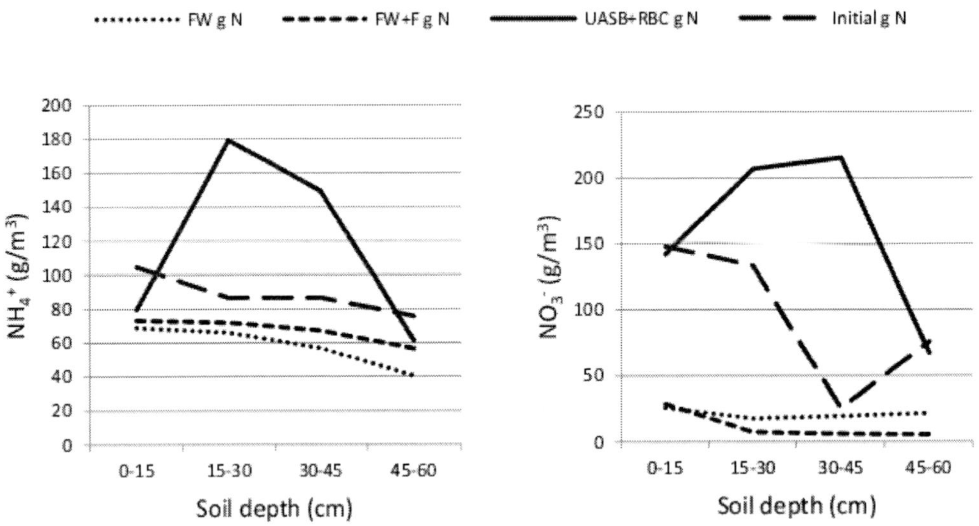

Figure 3. Speciation of ammonium (NH_4^+ left figure) and nitrate nitrogen (NO_3^- right figure) along the tested soil profile.

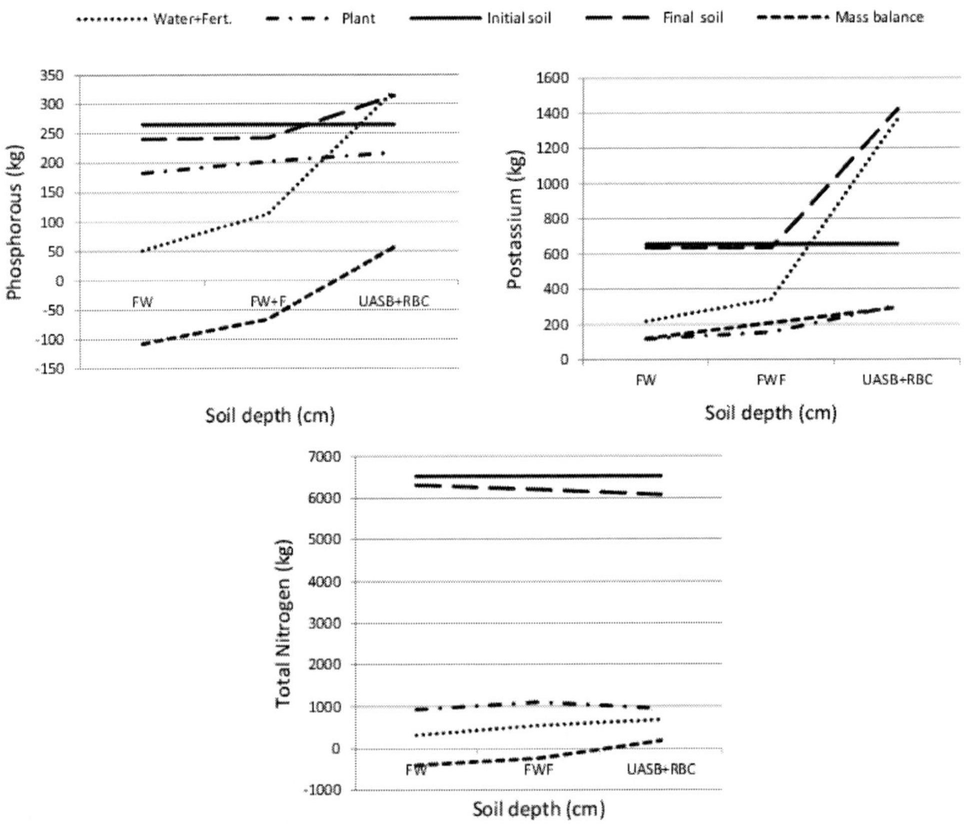

Figure 4. Phosphorous (top left), potassium (top right) and total nitrogen (bottom) mass balance in the top 30 cm of the 2.5m×2.5m test plots.

Therefore, negative balance values indicate that nutrients were leached to deeper layers with additional but less efficient utilization by the tomato plant as was also indicated in the previous section 4.2.

5. Pathogen Indicators

Table 3 list the pathogenic indicators analyzed for samples from the bottom, middle and top levels of the tomato fruits and leaves taken after two weeks from irrigation termination. The measured indicators from different levels were averaged to indicate the pathogenic quality measures on leaves and fruits. For the different sources of irrigation, *E.coli* and *T.coliform* was less than 1 MPN/100ml, which is within the limits of WHO guidelines (Table 1). The maximum counts of *Enterococcus* in samples from leaves were 240 MPN/100ml. Although the UASB-RBC sampled effluent contained much higher counts, the counts on the plants were much less and in an acceptable range. The maximum counts were mainly detected in sample from the lower parts of the plant and still within an acceptable range.

Table 4 lists the biological indicators results of soil samples taken at different depths. Generally, counts of pathogen indicators were reduced considerably in soil samples collected 5 to 10 days after the last irrigation compared to initial counts in the irrigation sources (Table 1). After 10 days of the last irrigation, all soil samples collected from all depths contained less than 1 MPN/100 ml of *E. coli*, less than 132 MPN/100 ml for *T. coliform* and less than 386 MPN/100 ml for *Enterococcus*. All these counts are within acceptable range compared to the WHO contents in Table 1. Comparing pathogenic indicators at different depths with respect to the three irrigation sources, after 10 days of the last irrigation, irrigation with UASB-RBC effluent give less counts of *E. coli* and *T. coliform* through the soil profile compared with FWF. After 5 days of the last irrigation, counts of *T.coliform* and *Enterococcus* indicators were higher in FW irrigated plots compared to test plots irrigated with UASB-RBC effluent, while the case is different for *E.coli* counts which shows higher counts in plots irrigated with UASB-RBC effluent. Such differences in pathogenic indicators indicate differences in both transport and decay rates of pathogens with respect to irrigation sources.

DISCUSSION

The preliminary decision of using a certain type of treated wastewater for irrigated agriculture and to take advantage of its nutrients values is based on quality, which is determined by its chemical and biological characteristics.

Table 3. Pathogenic indicators in tomato crop (leaves and fruits) for all treatments

Biological indicator	FW		FWF		UASB-RBC	
	Leaves	Fruits	Leaves	Fruits	Leaves	Fruits
E. coli (MPN/100ml)	<1	<1	<1	<1	<1	<1
T.coliform (MPN/100ml)	<1	<1	<1	<1	<1	<1
Enterococcus (MPN/100ml)	<1	<1	<1	<1	$<1\text{-}2.3\times10^3$	<1-20

Table 4. Pathogenic indicator concentrations for soil samples taken at different depths for all applied treatments after 5 to 10 days of last irrigation

E. coli (MPN/100ml)						
	FW		FWF		UASB-RBC effluent	
Soil depth	5 days	10 days	5 days	10 days	5 days	10 days
0-15	<1	<1	<1	<1	400	<1
15-30	63	<1	<1	<1	120	<1
30-45	<1	<1	<1	<1	<1	<1
45-60	160	<1	<1	<1	41	<1
T. coliform (MPN/100ml)						
	FW		FWF		UASB-RBC effluent	
Soil depth	5 days	10 days	5 days	10 days	5 days	10 days
0-15	5.79E+03	132	565	63	1.17E+03	<1
15-30	2.40E+04	<1	74	40	340	<1
30-45	97	<1	<1	<1	3.07E+03	<1
45-60	930	<1	41	<1	500	<1
Enterococcus (MPN/100ml)						
	FW		FWF		UASB-RBC effluent	
Soil depth	5 days	10 days	5 days	10 days	5 days	10 days
0-15	2.40E+04	382	8.66E+03	274	1.11E+04	146
15-30	1.55E+04	242	3.07E+03	30	1.30E+04	386
30-45	1.19E+03	20	930	30	660	119
45-60	530	0.0	1.78E+03	0.0	490	135

Sustainability of such reuse systems depends on keeping minimal health risks and environmental impacts while maintaining crop and energy yields and soil fertility and characteristics. Common environmental risks such as salinity, sodicity, nutrients, heavy metals and inorganic and organic contaminants, have been well documented (USEPA 2004; WHO 2006; Stevens 2006; Hamilton et al. 2007). These sustainability objectives are discussed in the following sections in the light of the obtained results from the test plots comparing reuse of UASB-RBC treated effluents with conventional use of fresh water with and without fertilizer in drip irrigated systems under typical Mediterranean arid climate in Jordan.

1. Decentralized System Productivity

Irrigation with UASB-RBC effluent resulted in higher dry weights of fruits, stems and leaves compared to FWF and FW treatments. Fresh tomato yields were also greater irrigating with UASB-RBC effluent, which means that UASB-RBC effluent have under their current conditions fertilizing capacity equal to the amount of nutrients added by the Jordanian farmer. The results indicate no major limitations to the use of UASB-RBC effluent as an irrigation source for tomato plant. As a matter of fact, the nutrient content of the UASB-RBC effluent was able to maintain good plant growth better than FWF which reflect the beneficial effect of UASB-RBC effluent as nutrients source that can supply relatively high contribution of tomato N, P and K requirements.

These results are in agreement other reported research that concluded that variable nutrient supply from treated wastewater depending on element, areas and method and level of treatment. Alkhamisi et al. (2011) reported that plants irrigated with treated wastewater had higher development rates and dry matter yield in comparison with those irrigated with fresh water as a result of available nutrients in the treated wastewater. They attributed the increase of plant height in comparison with the fresh water to higher amounts of nutrients, especially nitrogen in the treated wastewater.

Since no additional nutrients were added to any of the treatments, treated wastewater plots yielded better growth rates by virtue of their higher nitrogen and possibly sulfate contents compared with fresh water (Alkhamisi et al., 2011). Parsons et al. (2010) also reported the importance of nutrients, such as nitrogen, in reclaimed water and their positive effects on plant growth. This is similar to results reported by others (Duqqah, 2002, Qadir et al., 2010b and Evanylo et al., 2010). Irénikatché Akponikpè et al. (2011) found that the reuse of treated wastewater to irrigate tomato and eggplant significantly improved compared to that of the freshwater and treated wastewater provided variable nutrients supply depending on year and element.

The same effect was found on several vegetables in Morocco by Bouhoum and Amahmid, 2002 comparing three water irrigation sources (untreated wastewater, treated wastewater, surface water). They found that wastewater reuse increased yield by 13% for mint, 20.3% for coriander, 31.6% for radish and 35.1% for carrot. Similar results were also obtained in Soudan by Mirghani et al. (2002) on fodder crop irrigated with treated wastewater. They noticed a significant effect of wastewater irrigation on fodder crops growth, number of leaves, stem diameter, leaf area and aboveground biomass. Parsons et al. (2010) reported that high application rates of treated wastewater to citrus increased tree growth and fruit production.

However, there was insufficient macronutrient content in the treated wastewater to meet plant nutritional requirements. In a related study, treated wastewater has also been used for turf grass irrigation (Evanylo et al., 2010).

Moreover, Qadir et al. (2010a) reported that the use of treated wastewater has increased, as millions of small-scale farmers in urban areas of developing countries depend on treated wastewater or polluted water sources to irrigate high-value edible crops for urban markets. Concern was expressed about the harm to human health.

Mohammad and Ayadi (2004) noted that the uptake of macro and micronutrients by corn increased with treated wastewater irrigation, implying that secondary treated wastewater could be a source of plant nutrients and can be reused for irrigation to increase forage crop production. Tavassoli et al. (2010) reported a major increase in fresh and dry forage yield of corn irrigated with treated wastewater with a significant influence on crude protein content, ash percentage, and macro elements (i.e., N, P and K). Crop response to treated wastewater may be ascribed to water and nutrient supplies but mainly because nutrient are provided and released continuously (e.g. Feigin et al., 1991; Fonseca et al., 2007b).

In addition to productivity of crop, the observed methane yield can be utilized for power generation. For a typical implementation of the UASB for 5 households with average family size of 6.65 (JMEMR, 1997) and 127 l/capita/day potable water consumption (JMEMR, 1997.) that is typically recorded for the case study area, the observed biogas yield of 0.6 m^3 CH_4/m^3 wastewater gives a potential of 14.5 kWh/day power generation for the five households.

2. Maintenance of Soil Fertility

The application of treated wastewater in soil cultivated with tomato crop was efficient in supplying some essential nutrients for plant growth, such N, P, K, plus micronutrients promoting augments in concentration of these nutrients in soil, and also reductions in recommended doses of fertilizers, therefore, decreasing costs with fertilization.

UASB-RBC effluent revealed that the total K content of soil did not change in comparison with the very high initial K concentration. There was a slight increase in the total P content and a considerable increase in the TN (NH^{4+} and NO^{3-}) content of the soil. It is evident from the value of N, P, and K of treated wastewater applied that the N content is always in excess of crop requirement, which results in accumulation of N in soil. The other nutrients are not present in excess; hence, after the growth of tomato crop, the levels were reduced. The crops seem to remove P and K from the soil reserves, because the supply through irrigation with treated wastewater is inadequate.

In order to maintain the fertility of the soil with respect to P and K, nutrient enhancement of treated wastewater is necessary. Al Omron et al. (2012) on a study of long term use of wastewater in irrigation resulted in significant increase in organic matter content ranging from 17% to 30% in wastewater irrigated soils as compared to well water irrigated ones. Long term wastewater irrigation resulted in significant build-up of total concentration of Zn (130%), Pb (55%), Fe (82%), Ni (84%), Mn (30%), Cu (40%), Cr (75%), Co (78%) and As (67%) in wastewater irrigated soil samples over adjacent well water-irrigated soil samples. Mohammad and Mazahareh (2003) noted that treated wastewater irrigation increased soil phosphorus (P), potassium (K), iron (Fe), and manganese (Mn) levels, but soil organic matter was increased only in the topsoil.

3. Effects on Soil Physical and Chemical Properties

Although silty-clay-loam soil is difficult to till and manage, this soil type supplies plant nutrients and lime with high water holding capacity. The presence of surface plastic mulch covering the experimental area influences the soil moisture regime by controlling evaporation from the soil surface and thus increases water-use efficiency (Mamkagh, 2009; Cirelli et al., 2012). The soil is classified according to its possible use for irrigation (based on their potential to earn money from irrigated farming) as class 2 with intermediate suitability and payment capacity.

The pH value for UASB-RBC effluent was 7.78 fall within the pH range 6.5 – 8.0, which is suitable for irrigation under most conditions (FAO guidelines, Table 1). The initial level of soil pH at different depth ranged from 8.0-8.2 (Table 2), which poses no problem for use of treated wastewater in the area for irrigation development. High pH values indicate alkalinity problem.

Also, sodium ion is likely to be the dominant cation in the soil colloid, which can lead to flocculating soil condition, decline in aeration and poor internal drainage, resulting in water logging and therefore affecting normal plant developmental processes. Toxicity and availability of metals is reduced at neutral to alkaline pH since they will be insoluble (Foth et al., 1997). According to Pescod and Arar (1985), heavy metals are more available for plant uptake under acid conditions than in neutral to alkaline conditions as they are in ion form.

Pescod (1992) argues that a pH of <6.5 allows leaching of most metals in soluble state and pH > 10 causes movement of trace metals to be inhibited and distraction of bacteria may occur.

EC of UASB-RBC effluent was 2.1 dS/m (Table 1) and classified according to FAO guidelines as slight to moderate restriction on use for irrigated agriculture. Initial soil EC_e was ranging from 0.43 dS/m at the top layer to 0.32 dS/m at bottom layer and classified as non-saline soil with negligible salinity effect on crop growth (Abrol et al., 1988). The prolonged reuse of treated wastewater in irrigation will gradually be accompanied by salinity build up in the soils especially when proper management is not adopted. The particle size analysis, has however, shown that the soil made up of silty-clay-loam with high water holding capacity and low drainage which could result in soil salinization in the future, especially, if proper management practices such as the increase in frequency of application of irrigation water, good drainage system and soil leaching not practiced. Bouchaib et al. (1999) found that treated wastewater applications attenuated the detrimental effects of water salinity on crops. Feachem et al. (1982) observed that good drainage and proper management would lower the contact period between the plant and saline waters. The lethal concentration of total salts in irrigation water will result in soil salinization, which in turn may raise the osmotic pressure of the soil and so prevents the absorption of water by plants.

The sodium content of UASB-RBC effluent was 9.0 me/l, which has a severe restriction on use by sprinkler and surface irrigation (Ayers and Wescot, 1994). High amounts of sodium ions can result in the precipitation and granulation of calcium and magnesium ions from the soil thus affecting their effectiveness in enhancing physical internal drainage. The reuse of treated wastewater in irrigation may gradually be followed by adverse sodicity problems, which will have marked influence on soil permeability, resulting in a soil with poor internal drainage thus affecting infiltration of water. High amounts of total salts as indicated by electrical conductivity of the tested area can help to improve soil structure by the deflocculating action which, tends to counter the poor physical condition caused by high sodium concentration in the soil.

4. Health Risk Assessment

Drip irrigation applies the water direct to the roots with minimal transfer of pathogens to the crop surface (Pescod, 1992). Based on the results from the field experiments we found that drip irrigation is safer than surface and sprinkler irrigation as almost all tomato samples were free of pathogenic contamination. It was not possible to detect any reduced pathogenic contamination associated with drip irrigation, as only very few tomato samples were contaminated with *E. coli*, *T. coliform* and *Enterococcus* which are found to be for fruits and leaves collected from the bottom of tomato plants, even their values still below the recommended WHO limits (Table 1). Other studies have shown that wastewater reuse in irrigated vegetables grown above the soil surface contained higher levels of *fecal* contamination when surface-irrigated compared with those that were subsurface-irrigated (Sadovski et al., 1978; Assadian et al., 2005; Forslund et al., 2012). In a field study by Fonseca et al. (2011) comparing sprinkler, furrow and subsurface drip irrigation of lettuce, it was found, that sprinkler irrigation resulted in substantial higher concentrations of *E. coli* while almost all lettuce samples from subsurface drip irrigation were free of *E. coli*.

The irrigation water was inoculated with *E. coli* at a concentration of 10^8-10^9 cells ml^{-1} (Fonseca et al., 2011). Comparison of spray or surface irrigated lettuce grown in pots with irrigation water inoculated with E. coli O157:H7 demonstrated significantly higher contamination of lettuce (Solomon et al., 2002). In general it is expected that fruit and vegetable contamination will occur when surface irrigation is used, however, in Pakistan low levels of fecal contamination was found on vegetables irrigated through basin irrigation despite high concentrations of *E. coli* in the irrigation water (Ensink et al., 2007). These interesting results mean that the cost of wastewater treatment may decrease considerably when proper irrigation techniques and agricultural practices are applied. However, it is believed that with suitable irrigation systems, wastewater with much higher pathogenic counts could be used. Under these conditions, drip irrigation system and certain agricultural practices such as covering the soil with plastic mulch and wastewater treatment and human exposure control were chosen to minimize the contamination risk of farm workers and potential consumers of such crops.

Irrigation of tomatoes continued until harvest due to the hot climate at site of the experiment while irrigation was terminated two weeks before harvest at the green house site. The period with no irrigation could have reduced the concentration of *E. coli* on tomatoes crop below the detection level. The time interval between the last irrigation event and harvest determines the extent of contamination, as pathogens have been shown to decline with time following cessation of irrigation before harvest. It is recommended that irrigation water for tomatoes is of an appropriate microbial quality particularly when applied close to the time of harvest or during harvesting to avoid contamination of the tomatoes (FDA, 2009). Tomato and other crop growers should maximize the time between the last irrigation event and harvest to minimize food safety risks when irrigating with water of low quality is unavoidable (Barker-Reid et al., 2009), e.g. termination of overhead irrigation seven days before crop harvest has been recommended (Rangarajan et al., 2000).

The contamination of vegetables grown in soil irrigated with treated wastewater will largely depend on the survival capabilities of the pathogens in the soil and on plants. A common maximum survival time reported for bacterial pathogens in soil is two months (Gerba and Smith, 2005), but persistence of bacterial pathogens in soil may be up to five months due to moist conditions and protection against sunlight (Islam et al., 2004). A considerable longer survival time, e.g. up to two years, has been reported for helminth eggs in soil (Gerba and Smith, 2005). Soil samples in the present study were collected after 5 and 10 days of irrigation termination. After 10 days of last irrigation, *E. coli* count did not exceed 1 MPN/100 ml in all soil samples, which shows a short survival time for *E. coli* after irrigation. Similar results were also reported by other researchers (Entry et al., 2005) who did not detect *E. coli* in any soil samples after the first day of dairy manure and compost application to soil. The count of total coliforms decreases with time and soil dryness as shown after 10 days of irrigation cessation. With respect to *Enterococcus*, there was reduction in soil samples taken at all depth after five days of the last irrigation compared with counts present in irrigation wastewater. It should also be noted that *E. coli*, *T. coliform* and *Enterococcus* were mainly found in the upper soil fraction both after 5 and 10 days of last irrigation in control plots (FW) and plots of FWF. This indicates that although the treated wastewater could have been a source of contamination, the very high concentration of pathogenic indicators found in soil samples (higher than pathogenic indicators in soil samples irrigated with UASB-RBC effluent) suggests that an external environmental source of *fecal* contamination would be more likely. This is supported by findings in Crete where soil samples from both tap water and secondary treated wastewater irrigated plots contained *E. coli* at harvest time (Forslund et al., 2012). Birds, insects, wild and domestic animals have been reported as sources of *fecal* contamination of the environment and fresh produce (Beuchat, 1996; Ishii and Sadowsky, 2008; Talley et al., 2009). Wild birds may harbor many pathogens including *E. coli* O157:H7 (Hubãlek, 2004). A study using library-dependent and library-independent microbial source tracking of *E. coli* showed birds and wastewater to be responsible for the *fecal* contamination of a beach (Edge et al., 2010).

The survival of these pathogens in soil layers is expected to decrease with time or soil dryness; however, with suitable environmental and soil conditions as humidity could help in growth of enteric bacteria again in the soil as reported by several authors (Entry et al., 2000 and Rufete et al., 2006). Moreover, protection of public health cannot be achieved solely at the treatment plant, as regrowth of pathogens has been reported, especially when chlorine is used for disinfection (Tchobanoglous et al., 2003). Instead, health protection can be achieved by using a multiple barriers approach (Carr et al., 2004) that interrupt the flow of pathogens to human. Health education will greatly improve awareness on possible health hazard in wastewater use as Blumethal et al. (2000) suggested where protection measures as health education to farmers about avoiding direct contact with wastewater and importance of hand washing using soap after contact with wastewater could determine the fate of pathogens in wastewater.

Wastewater is a resource that should be fully exploited and irrational standards will limit the use of this resource. In the decision-making process, guidelines for water use developed by one country may not be suitable for another due to the limitations such as technology and economic status (Asano et al. 2007). Without compromising the safety of organisms within the receiving system, guideline values may require adjustment based on case-specific treatment goals and the available technology. Many countries have adopted and/or modified water quality guidelines outlined by WHO (1989). Recently, the WHO has made modifications to their proposed guidelines for the reuse of water in agriculture based on the findings from epidemiological studies and quantitative microbial risk assessments (Brissaud 2008). To determine water quality guidelines, the WHO takes into account the cost of water treatment prior to reuse as well as health risks (Asano et al. 2007). Both cost and health risks largely determine the potential beneficial use of the treated water. Because of these factors, the guideline values from the WHO (2006) may be less stringent than values from other sources.

CONCLUSION

Water and clean water in particular, is scarce in Jordan. The lack of sufficient water resources is endangering the country's economic growth, as two of the main economic pillars, agriculture and tourism, are largely dependent on the availability of sufficient water of good quality. Traditional wastewater treatment technologies are expensive because of the high investment costs, high operational costs (electricity costs in particular) and high maintenance costs. The application of UASB technology will lead to a considerable reduction of these investments and to substantially lower operational costs.

The traditional treatment plants deliver drinking water quality. The UASB plant will deliver irrigation water quality with the possibilities of water and nutrients utilization, and as compared to aerobic mainstream technologies, does not require energy for oxygen input; on the contrary it enables energy recovery from organic matter and thus less vulnerable.

So UASB is proposed at a comparably very low cost to deliver just a little less clean water and is widely recognized as a core component in sustainable decentralized wastewater treatment schemes.

Decentralized management of wastewater facilitates utilization of wastewater as a valuable resource, instead of a pollutant, which when properly managed can provide a source of locally produced reusable water, fertilizer, and energy, in addition to improvements in public health and sanitation. Management of decentralized systems to incorporate concepts that enhance the value of wastewater provides a sustainable management solution, which delivers positive economical, social, and environmental benefits to communities.

Results of this study show that tomato yield increased when treated wastewater is used showing that treated wastewater can be used as an alternative water resource in water scarcity, especially in arid and semi-arid regions, to satisfy consumers' food demand. In addition to this advantage, it is also an organic fertilizer that may improve plant growth, reduce fertilizer application or reduce mineral fertilization cost and increase productivity of poorly fertile soils.

The main reason for concern in using treated wastewater is the presence of pathogens as they have the potential to cause disease in susceptible consumers. The microbial analysis of tomato fruits and leaves reveals no to slight accumulation of pathogenic indicators with *E. coli* and *T. coliform* concentrations of <1 MPN/100ml which shows that treated wastewater can be used to irrigate tomato or other vegetables that are eaten uncooked or cooked.

Treated wastewater irrigation can be used as an organic fertilizer to improve the physical and chemical properties of soils as a potential source of nutrients.

The major disadvantage of treated wastewater irrigation is the accumulation of immobile heavy metals in the soil. Contamination with heavy metals should be further investigated in order to determine the residual effects of treated wastewater before using it for land reclamation and as a fertilizer.

Use of treated wastewater for irrigation has gained importance throughout the world due to limited water sources and costly wastewater treatment for discharge. If land with suitable topography, soil characteristics and drainage is available, treated wastewater effluent can be put to good use as a source of both irrigation water and plant nutrients.

Recommendations:

1 Further studies should be implemented to test the long-term impact of wastewater irrigation on soils and crops.
2 According to these preliminary results a large-scale application of reclaimed wastewater is realistic.
3 Modeling pathogen and nutrients movement in soil with time and space should take place in order to predict possible soil and groundwater pollution.

ACKNOWLEDGMENTS

This research was part of the activities carried out in a CORETECH-EU, ENCOMED-funded project: The development of cost-effective reclamation for domestic wastewater and the appropriate reuse of treated effluent under semi-arid climate conditions. The author would like to acknowledge the J. William Fulbright Scholarship Board (FSB) for a Fulbright Grant as visiting scholar at Washington State University, US.

REFERNCES

Abrol, I. P., Yadav, J. S. P., Massoud, F. I. (1988). *Salt-Affected Soils and their Management: Soils bulletin 39*. Rome: FAO.

Al Omron, A. M., El-Maghraby, S. E., Nadeem, M. E., A. El-Eter, A. M., Al-Mohani, H. (2012). Long term effect of irrigation with the treated sewage effluent on some soil properties of Al-Hassa Governorate, Saudi Arabia. *Journal of the Saudi Society of Agricultural Sciences*, 11, 15–18.

Alkhamisi, S. A., Abdelrahman, H. A., Ahmed, M., Goosen, M. F. A. (2011). Assessment of reclaimed water irrigation on growth, yield, and water-use efficiency of forage crops. *Applied Water Science*, 1, 57–65

APHA, AWWA, WPCF. (1998). *Standard Methods for the Examination of Water and Wastewater: 20th ed.* NW Washington, 2015 Fifteenth Street, DC 20005: American Public Health Association, American Water Works Association and Water Environment Federation.

Asano, T., Burton, F. L., Leverenz, H., Tsuchihashi, R., Tchobanoglous, G. (2007). *Water Reuse: Issues, Technologies, and Applications*. New York: McGraw-Hill.

Assadian, N. W., Di Giovanni, G. D., Enciso, J., Iglesias, J., Lindemann, W. (2005). The transport of waterborne solutes and bacteriophage in soil subirrigated with a wastewater blend agriculture. *Ecosystems and Environment*, 111 (1-4), 279-291.

Ayers, R. S., Wescot, D. W. (1994). *Water Quality for Agriculture: Irrigation and Drainage Paper 29*. Rome: FAO.

Barker-Reid, F., Harapas, D., Engleitner, S., Kreidl, S., Holmes, R., Faggian, R. (2009). Persistence of Escherichia coli on injured iceberg lettuce in the field overhead irrigated with contaminated water. *Journal of Food Protection*, 72 (3), 458-464.

Beuchat, L. R. (1996). Pathogenic microorganisms associated with fresh produce. *Journal of Food Protection*, 59 (2), 204-216.

Blumenthal, U. J., Duncan Mara, D., Peasey, A., Ruiz-Palacios, G., Stott, R. (2000). Guidelines for the microbiological quality of treated wastewater used in agriculture: recommendations for revising WHO guidelines. *World Health Orgnaization*, 78, 1104–1116.

Bouchaib, E. H., Handoufb, A., Mekraneb, M., Touzanib, M., Khanac, A., Khallayounea, K., Benchokround, T. (1999). Use of wastewater for crop production under arid and saline conditions: yield and hygienic quality of the crop and soil contaminations. *Water Science Technology,* 33:10/11, 327–334.

Bouhoum, K., Amahmid, O. (2002). Municipal wastewater reuse for irrigation: productivity and contamination level of irrigated crops by pathogens: *International Symposium on Environmental Pollution Control and Waste Management, EPCOWM*, Tunis.

Brissaud, F. (2008). Criteria for water recycling and reuse in the Mediterranean countries. *Desalination*, 218, 24–33.

Bruinsma, J. (2009). The Resource Outlook to 2050: By How Much do Land, Water and Crop Yields Need to Increase by 2050? *FAO Expert Meeting on 'How to Feed the World in 2050'*. Rome: FAO.

Carr, R. M., Blumenthal, U. J., Mara, D. D. (2004). *Health guidelines for the use of wastewater in agriculture: developing realistic guidelines: Wastewater Use in Irrigated Agriculture* (Scott, C. A., Faruqui, N. I., Raschid-Sally, L. (Eds.)). Wallingford : CABI Publishing.

Cirelli, G. L., Consolia, S., Licciardello, F., Aiello, R., Giuffridab, F., Leonardib, C. (2012). Treated municipal wastewater reuse in vegetable production. *Agricultural Water Management*, 104, 163– 170.

Crites, R., Tchobanoglous, G. (1998). *Small and Decentralized Wastewater Management Systems*. New York: WCB/McGraw-Hill.

Duqqah, M. M. (2002). Treated sewage water use in irrigated agriculture: theoretical design of farming systems in Seil Al Zarqa and Middle Jordan Valley in Jordan. *The Netherlands: Published Ph.D. thesis of the Wageningen University*.

Edge, T. A., Hill, S., Seto, P., Marsalek, J. (2010). Library-dependent and library independent microbial source tracking to identify spatial variation in faecal contamination sources along a Lake Ontario beach (Ontario, Canada). *Water Science Technology*, 62, 719–727.

Elmitwalli, T. A., Sayed, S., Groendijk, L., van Lier, J., Zeeman, G., Lettinga, G. (2003). Decentralised treatment of concentrated sewage at low temperature in a two-step anaerobic system: two upflow-hybrid septic tanks. *Water Science Technology*, 48(6):

Ensink, J. H. J., Mahmood, T., Dalsgaard, A. (2007). Wastewater irrigated vegetables: market handling versus irrigation water quality. *Tropical Medicine and International Health*, 12 (2), 2-7.

Entry, J. A., Sojka, R. E. (2000). The efficacy of polyacrylamide related compounds to remove microorganisms and nutrients from animal wastewater. *Journal of Environmental Quality*, 29, 1905–1914.

Entry, L. A., Leytem, A. B., Verwey, S. (2005). Influence of solid dairy manure and compost with and without alum on survival of indicator bacteria in soil and on potato. *Environmental Pollution*, 138, 212-218.

Evanylo, G., Ervin, E., Zhang, X. (2010). Reclaimed water for turfgrass irrigation. *Water* 2, 685–701.

Feachem, R. G., Bradley, D. J., Garelick, H., Mara, D. D. (1983). Reuse of excreta and discharge of effluents: Chapter 7: Health aspects of excreta and wastewater management. *Sanitation and Disease* (Eds.). Washington, D. C.: Wiley Sons.

Feigin, A., Ravina, I., Shalhevet, J. (1991). *Management for Environmental Protection: Irrigation with Treated Sewage Effluent.* Berlin: Springer-Verlag.

Fonseca, A. F., Herpin, U., Paula, A. M., Victria, R. L., Melfi, A. J. (2007). Agricultural use of treated sewage effluents: Agronomic and environmental implications and perspectives for Brazil. *Science Agriculture*, 64, 194–209.

Forslund, A., Ensink, J. H. J., Markussen, B., Battilani, A., Psarras, G., Gola, S., Sandei, L., Fletcher, T., Dalsgaard, A. (2012). Escherichia coli contamination and health aspects of soil and tomatoes (Solanum lycopersicum L.) subsurface drip irrigated with on-site treated domestic wastewater. *Water Research*, 46, 5917 -5934.

Foth, D. H., Ellis, B. (1997). *Soil Fertility* (second edition). US: Lewis Publishers.

Gerba, C. P., Smith, J. E. (2005). Sources of pathogenic microorganisms and their fate during land application of wastes. *Journal of Environmental Quality*, 34 (1), 42-48.

Guest, S. J., Skerlos, J. S., Barnard, L. J., Beck, B. J., Daigger, T. G., Hilger, H., Jackson, J. S., Karvazy, K., Kelly, L., Macpherson, L., Mihelcic, R. J., Pramanik, A., Raskin, L., Van Loosdrecht, M. C. M., Yeh, D., Love, N. G. (2009). A new planning and design paradigim to achieve sustainable resource recovery from wastewater. *Environmental Science Technology,* 43 (16), 6126–6130.

Hamilton, A. J., Stagnitti, F., Xiong, X., Kreidl, S. L., Benke, K. K., Maher, P. (2007). Wastewater irrigation: the state of play. *Vadose Zone Journal*, 6, 823–840.

Hubálek, Z. (2004). An annotated checklist of pathogenic microorganisms associated with migratory birds. *Journal of Wildlife Diseases*, 40 (4), 639-659.

Irénikatché Akponikpè, P. B., Wima, K., Yacouba, H., Mermoud, A. (2011). Reuse of domestic wastewater treated in macrophyte ponds to irrigate tomato and eggplant in semi-arid West-Africa: Benefits and risks. *Agricultural Water Management*, 98, 834–840.

Ishii, S., Sadowsky, M. J. (2008). Escherichia coli in the environment: implications for water quality and human health. *Microbes and Environments*, 23 (2), 101-108.

Islam, M., Doyle, M. P., Phatak, S. C., Millner, P., Jiang, X. (2004). Persistence of enterohemorrhagic Escherichia coli O157:H7 in soil and on leaf lettuce and parsley grown in fields treated with contaminated manure composts or irrigation water. *Journal of Food Protection*, 67 (7), 1365-1370.

Jamieson, R. C., Gordon, R. J., Sharpies, K. E., Stratton, G. W., Madani, A. (2002). Movement and persistence of fecal bacteria in agricultural soils and subsurface drainage water: a review. *Canadian Biosystem Engineering*, 44, 11–19.

Jordanian Ministry of Energy and Mineral Resources (JMEMR Report), Survey of Energy Consumption in Residential Sector. (1997). Amman, Jordan.

Lettinga, G., van Velsen, A. F. M., Hobma, S. W., De Zecuw, W., Klapwijk, A. (1980). Use of the upflow sludge blanket (USB) reactor concept for biological wastewater treatment, especially for anaerobic treatment. *Biotechnology Bioengineering*, 22, 699–734.

Lettinga, G., Roersma, R., Grin, P. (1983). Anaerobic treatment of raw domestic sewage at ambient-temperatures using a granular bed UASB reactor. *Biotechnology. Bioengineering*, 25 (7), 1701–1723.

Lettinga, G., deMan, A., van der Last, A. R. M., Wiegant, W., van Knippenberg, K., Frijns, J., (1993). Anaerobic treatment of domestic sewage and wastewater. *Water Science Technology*, 27(9), 67–73.

Lettinga, G. (2010). *Environmental anaerobic technology: application and new development* (Fang, H, H, P. Ed.), Chapter 1:The route of anaerobic waste (water) treatment toward global acceptance. 57 shelton street, London: Imperial College Press.

Levine, A. D., Asano, T. (2004). Recovering sustainable water from wastewater. *Environmental Science Technology*, 38 (11), 201A–208A.

Mahmoud, N. (2008). High strength sewage treatment in a UASB reactor and an integrated UASB-digester system, *Bioresources Technology*, 99, 7531–7538.

Mamkagh, A. M. (2009). Effect of tillage and plastic mulch on growth and yield of Okra (Abelmoschus esculentus) grown under rainfed conditions. *International Agricultural Biology,* 11, 453–457.

Mirghani, T. I., Babikir, H. N., Nasre, M. M. (2002). Utilization of treated sewage water for forage production, In: *International Symposium on Environmental Pollution Control and Waste Management, EPCOWM*, Tunis.

Mohammad, M. J., Mazahreh, N. (2003). Changes in soil fertility parameters in response to irrigation of forage crops with secondary treated wastewater. *Communications in Soil Science and Plant Analysis,* 34 (9/10), 1281–1294.

Mohammad, M. J., Ayadi, M. (2004). Forage yield and nutrient uptake as influenced by secondary treated wastewater. *Journal of Plant Nutrition*, 27(2), 351–365.

Oktem, Y. A., Ince, O., Sallis, P., Donnelly, T., Ince, B. K. (2008). Anaerobic treatment of a chemical synthesis-based pharmaceutical wastewater in a hybrid upflow anaerobic sludge blanket reactor. *Bioresources Technology*, 99, 1089–1096.

Parsons, L. R., Sheikh, B., Holden, R., York, D. W. (2010). Reclaimed water as an alternative water source for crop irrigation. *Horticulture Science*, 45(11),1626–1629.

Pescod, M. B. (1992). Wastewater Treatment and Use in Agriculture. *Irrigation and Drainage Paper 47*. Rome: FAO.

Pescod, M. B., Arar, A. (1985). *Treatment and use of Sewage Effluent for Irrigation: Proceedings of the FAO Regional Seminar on the Treatment and Use of Sewage Effluent for Irrigation held in Nicosia*, Cyprus: FAO.

Qadir, M., Bahri, A., Sato, T., Al-Karadsheh, E. (2010a). Wastewater production, treatment, and irrigation in Middle East and North Africa. *Irrigation and Drainage Systems*, 24, 37–51.

Qadir, M., Wichelns, D., Raschid-Sally, L., McCornick, P. G., Drechsel, P., Bahri, A., Minhas, P. S. (2010b). The challenges of wastewater irrigation in developing countries. *Agricultural Water Management*, 97 (4), 561–568.

Rangarajan, A., Bihn, E. A., Gravani, R. B., Scott, D. L., Pritts, M. P. (2000). Food Safety Begins on the Farm: a Grower's Guide. Available at: http://ecommons.cornell.edu/handle/1813/2209.

Rebhun, M. (2004). Desalination of reclaimed wastewater to prevent salinization of soils and groundwater. *Desalination*, 160, 143–149.

Rufete, B., Perez-Murcia, M. D., Perez-Espinosa, A. P., Moral, R., Moreno-Caselles, J., Paredes, C. (2006). Total and faecal coliform bacteria persistence in a pig slurry amended soil. *Livestock Science*, 102, 211 – 215.

Ryan, J., Garabet, S., Harmsen, K., Rashid, A. (1996). *A Soil and Plant Analysis Manual for the West Asia and North Africa Region*. Aleppo, Syria: ICARDA.

Sadovski, A. Y., Fattal, B., Goldberg, D. (1978). Microbial contamination of vegetables irrigated with sewage effluent by the drip method. *Journal of Food Protection*, 41 (5), 336-340.

Soil Survey Staff (2011). Soil Survey Laboratory Information Manual: Burt. R. (Eds.). *Soil Survey Investigations Report No. 45, Version 2.0.* US: US Department of Agriculture, Natural Resources Conservation Service.

Solomon, E. B., Potenski, C. J., Matthews, K. R. (2002). Effect of irrigation method on transmission to and persistence of Escherichia Coli O157:H7 on lettuce. *Journal of Food Protection*, 65(4), 673–676.

Stevens, D. (Ed.) (2006). *Growing crops with reclaimed wastewater*. Collingwood: CSIRO.

Talley, J. L., Wayadande, A. C., Wasala, L. P., Gerry, A. C., Fletcher, J., DeSilva, U., Gilliland, S. E. (2009). Association of Escherichia coli O157:H7 with filth flies (Muscidae and Calliphoridae) captured in leafy greens fields and experimental transmission of E. coli O157:H7 to spinach leaves by house flies (Diptera: Muscidae). *Journal of Food Protection*, 72 (7), 1547-1552.

Tandukar, M., Uemura, S., Ohashi, A., Harada, H. (2005). A low-cost municipal sewage treatment system with a combination of UASB and the "Fourth Generation" down-flow hanging sponge (DHS) reactors. *Water Science Technology*, 52(1–2), 323–9.

Tarchitzky, J., Lerner, O., Shani, U., Arye, G. L. A. A., Brener, A., Chen, Y. (2007). Water distribution pattern in treated wastewater irrigated soils: hydrophobicity effect. *European Journal of Soil Science*, 58(3), 573 – 588.

Tauseef, S. M., Abbasi, T., Abbasi, S. A. (2013). Energy recovery from wastewaters with high-rate anaerobic digesters. *Renewable and Sustainable Energy reviews*, 19, 704-741.

Tavassoli, A., Ghanbari, A., Amiri, E., Paygozar, Y. (2010). Effect of municipal wastewater with manure and fertilizer on yield and quality characteristics of forage in corn. *African Journal of Biotechnology,* 9(17), 2515–2520.

Tawfik, A., Zeeman, G., Klapwijk, A., Sanders, W., El-Gohary, F., Lettinga, G. (2003). Treatment of domestic sewage in a combined UASB/RBC system. Process optimization for irrigation purposes. *Water Science Technology*, 48(1), 131–8.

Tchobanoglous, G. ,Burton, F. L., Stensel, H. D. (2003). *Wastewater Engineering: Treatment and Reuse* (fourth ed.). New York: McGraw Hill.

Travis, M. J., Wiel-Shafran, A., Weisbrod, N., Adar, E., Gross, A. (2010). Greywater reuse for irrigation: effect on soil properties. *Science Total Environment* 408, 2501–2508.

United Nations Department of Economic and Social Affairs, Population Division (UNDESA). (2009a). World Population Prospects: The 2008 Revision, Highlights, Working Paper No. ESA/P/WP.210. New York: UN.

US Environmental Protection Agency (USEPA). (2004). Guidelines for water reuse. US Agency for International Development. Available at: http://www.epa.gov/nrmrl/pubs/625r04108/625r04108.htm.

US Environmental Protection Agency (USEPA). (2011). Global anthropogenicnon-CO2 greenhouse gas emissions: 1990–2030. United State Environmental Protection. Available at: /http://www.epa.gov/climatechange/economics/downloads/EPA_NonCO2_Projections_2011_draft.pdfS.

US Food and Drug Administration (FDA). Guidance for Industry: Guide to minimize microbial food safety hazards of tomatoes; Draft Guidance. (2009). Available at: http://www.fda.gov/Food/GuidanceComplianceRegulatoryInformation/GuidanceDocuments/ProduceandPlanProducs/ucm173902.htm.

Verbyla, E. M., Oakley, S., Mihelcic, J. (2013). Wastewater infrastructure for small cities in an urbanizing world: Integrating the protection of human health and the environment with resource recovery and food security. *Environmental Science Technology*. Just Accepted, DOI: 10.1021/es3050955.

Wallach, R., Ben-Arie, O., Graber, E. R. (2005). Soil water repellency induced by long-term irrigation with treated sewage effluent. *Journal of Environmental Quallity*, 34, 1910–1920.

World Health Organization (WHO). (1989). Health guidelines for the use of wastewater in agriculture and aquaculture. *Tech. Rep. No. 778*. Geneva: WHO.

World Health Organization (WHO). (2006). *Guidelines for the Safe Use of Wastewater, Excreta and Greywater: Wastewater Use in Agriculture vol. 2*. Geneva: WHO.

World Water Assessment Program (WWAP). (2012). *United Nations World Water Development Report 4: Facts and Figures, Managing Water under Uncertainty and Risk.* Paris/London :UNESCO Publishing.

World Water Assessment Program (WWAP). (2009). *United Nations World Water Development Report 3: Water in a Changing World.* Paris/London: UNESCO Publishing/ Earthscan.

Yetilmezsoy, K., Sakar, S. (2008). Development of empirical models for performance evaluation of UASB reactors treating poultry manure wastewater under different operational conditions. *Journal of Hazard Materials*, 153, 532–543.

In: Anaerobic Digestion
Editor: Ana Torales

ISBN: 978-1-62808-884-7

Chapter 4

EFFECTS OF SUBSTRATE-TO-INOCULUM RATIO ON ANAEROBIC DIGESTION OF VEGETABLE WASTES

Saraven Koothan[1], Ackmez Mudhoo[1,*] and Romeela Mohee[2]
[1]Department of Chemical & Environmental Engineering, Faculty of Engineering, University of Mauritius, Réduit, Mauritius
[2]National Research Chair in Solid Waste Management, Mauritius Research Council, Ebène, Mauritius

ABSTRACT

The effects of substrate to inoculum ratio (SIR) on the anaerobic digestion of vegetable wastes slurries were investigated using biochemical methane potential (BMP) assays. Cow dung was used as inoculum for the BMP assays. The vegetable wastes consisted of a mixture cabbage tops, cauliflower leaves, carrots, beetroots, tomatoes and potatoes which was ground to form a slurry and characterized to yield a total solids (TS) of 6.40%, volatile solids (VS) of 76.92%, chemical oxygen demand (COD) of 11870.50mg/L, soluble COD (sCOD) of 3600.00 mg/L and a pH of 6.00. The different SIR was 1:1 and 3:1. 2 other set-up was done for 100% inoculum and 100% substrate. Water was added so as to have a TS of 8%. Digestion proceeded in a 2L reactor with a headspace of 500mL kept for storage of biogas.

The TS, VS, pH, COD, sCOD, volatile fatty acids (VFA) were monitored for 34 days and the biogas production rates were monitored over a period of 60 days to assess the effects of SIR on the anaerobic digestion process chemistry. From the experiments, it was found that the pH varied from 5.72 to 9.05, VFA from 0 meq/L to 336.30 meq/L, COD from 3597.12mg/L to 18000mg/L and sCOD from 719.42mg/L to 5000mg/L. Highest VS reduction was achieved by reactor B (SIR 1:1). Volume of biogas produced for each BMP was as follows: 21200 ml for SIR 1:1, 18500 ml for SIR 3:1, for 100% inoculum we noted 1150 ml and for 100% substrate is 13050 ml of biogas. Results have shown that after 30 days, the BMP assay consisting of 50% inoculums yielded the highest cumulative production of biogas. However in the long run (after 60 days), an SIR ratio of 3:1 resulted in the highest biogas production rate.

---

[*] Corresponding Author: Cellphone: +2307584279, Office phone: +2304037772; Fax: +2304657144. Email: ackmezchem@yahoo.co.uk.

Keywords: Substrates, inoculum, vegetable waste, biogas, anaerobic digestion

INTRODUCTION

Energy is one of the most important factors to global prosperity but its dependency on fossil fuels as a primary energy source has led to global climate change, environmental degradation, and human health problems. In addition, the recent rise in the prices of oil and natural gas has shifted the current economy to alternative energy sources such as biogas (Sunaro et al., 2010). AD is a technology commonly used for the treatment of organic waste to yield biogas as a resourceful by-product (Sunarso et al., 2010; Verma, 2002). AD can also be defined as an attractive waste treatment practice in which both pollution control and energy recovery can be achieved (Chen et al., 2007). Furthermore, in the AD process the organic carbon is converted to its most oxidized (CO_2) and most reduced (CH_4) states (Raposo et al., 2009). AD is also a method in which waste treatment can be achieved, while at the same time, renewable energy in the form of methane and fertilizer in the form of digested residues (biosolids) can be recovered (Azadeh et al., 2009). The digestion process is broken down into four processes and these are detailed below:

1. Hydrolysis

Hydrolysis is the first stage in the AD process where long chains of the complex carbohydrates, proteins and lipids of organic matter (OM) are broken down into smaller units (Gerardi, 2003). Hydrolysis reactions are catalysed by extracellular enzymes called hydrolases (Juanga, 2005) and the hydrolysis rate is dependent on substrate and bacterial concentration as well as on environmental factors like temperature and pH (Brown, 2001).

2. Acidogenesis

Acidogenesis is the stage where the reduced compounds (the products of hydrolysis) are converted to a mixture of short chain molecules including volatile organic acids, alcohols, ketones, aldehydes, water, formate, acetate, carbon dioxide CO_2 and hydrogen H_2 by the action of acid forming (fermentative) bacteria known as acidogens (Li et al., 2010; Juanga, 2005).

3. Acetogenesis

In this step, the products of the acidogenesis process, that is the fatty acids, are converted into acetic acid, formate, CO_2 and H_2. From these products, methane can be produced by the methanogenic bacteria (Juanga, 2005).

4. Methanogenesis

In this last stage of the anaerobic process, all the intermediate products formed from the previous stages including acetic acid, H_2 and CO_2 as well as other substrates of which formic acid and methanol being the most important, are converted into methane and CO_2. (Juanga, 2005). Methanogenesis is the most sensitive and important reaction in the production of the biogas and are achieved by the action of Methanogens comprising of *methanobacterium, methanobacillus, methanococcus and methanosarcina.* Methanogens can be further divided into two groups namely the acetate consumers that undergo a process known as aceticlastic methanogenesis, and the hydrogen/carbon dioxide consumers which undertake a process of hydrogenotrophic methanogenesis.

Factors Affecting AD

Within the anaerobic digester, there are different parameters affecting the rates of the different steps of the digestion process such as the operating pH, oxidation and reduction potentials, temperature, nutrients and trace metals.(Appels et al., 2008; Khanal, 2008).

Operating pH

There are two groups of bacteria in terms of pH optima. Methanogens (methane-producing bacteria) are extremely sensitive to pH, performing best at an optimum pH between 7.8 and 8.2 while the acidogens (acid-producing bacteria) prefer a pH range between 5.5 and 6.5 (Appels, 2008; Khanal, 2008). According to Khanal (2008), in an environment where both the methanogens and acidogens coexist, the pH is kept within a range of 6.8 and 7.4 since methanogenesis is regarded as the rate-limiting step.

Oxidation and Reduction Potentials (ORP)

Morris (1975) reported that for the proper growth of the required anaerobes in any medium, it is important to maintain the ORP value within the range of -200 to -350mV at pH 7. Reducing environment is essential for the performance of methanogens (Archer and Harris, 1986; Hungate, 1969). Along with lowering the risk of radial generation, the absence of oxygen also ensures that Fe is maintained in its reduced state of Fe(II) and are not available in form of free ions to induce toxicity within the microbial cells (Edson et al., 2004).

Temperature

Temperature plays an important part in the physicochemical properties of the components found in the substrate by affecting the growth rate and metabolism of the micro-organisms and hence affecting the population dynamics in the anaerobic digester (Appels, 2008). The highest temperature efficiency for anaerobic conversion of organic matter (OM) has been reported as being 35-40^{0}C for mesophilic conditions and about 50^{0}C for thermophilic conditions. According to Appels (2008) however, the application of high temperatures (thermophilic) has a negative effect on the digestion process and it leads to an increase in the fraction of free ammonia, playing an inhibiting role for the microorganism. To avoid disruption in the microbial system, anaerobic processes should be preferably operated within a temperature range of 10-45^{0}C (Khanal, 2008; Appels, 2008). It is also important to maintain a stable operating temperature in the digestor since sharp and frequent fluctuations of temperature may affect the bacteria, especially the methanogens (Khanal, 2008).

Nutrients and Trace Metals

Though the involvement of nutrients and trace metals are not directly linked to waste stabilization, all microbial processes require these constituents as they are those essential components that are required for the growing of existing microbial cells and production of new cells; they provide a suitable physicochemical condition for the optimum growth of microorganisms. Consequently, if the waste stream does not have one or more of the required nutrients and trace elements, the waste degradability will be severely affected by the inability of the microbial cell to grow and produce new cells (Khanal, 2008).

Substrate to Inoculum Ratio (SIR)

AD offers practical solutions for the treatment of high strength organic wastes such as municipal solid waste (Ismail et al., 2007), sewage sludge (Appels et al., 2008), livestock wastewater (Cantrell et al., 2008) and food waste (Kim et al., 2007). The substrate used in this study is vegetable waste as it is produced in large quantities and has a high biodegradability (Azadeh et al., 2009) due to high moisture content of around 48% (Mohee, 2002) which ease their biological treatment (Bouallagui et al., 2003). Anaerobic digestion is a complex process requiring the presence of different types of microorganisms. For proper degradation, not only the addition of a balanced active innoculum containing all the necessary bacteria is essential, the amount in which the innoculum is added is important as well (Juanga 2005; Angelidaki, 2002).

The source of innoculum is a very important operational parameter, being a crucial concern in the selection of SIR as well as in the assessment of anaerobic biodegradability of solid wastes (Binod, 2008). The performance and stability of biomethanogenesis during AD is also influenced by the quality and quantity of the inoculums. Lowering the levels of inoculums may cause an imbalance to occur due to the rapid increase in the growth rate of acid forming bacteria (compared to methanogens). Moreover, when inoculum is exposed to toxic compound or environmental stress factor such as abnormal temperature and heavy metal, it gets affected during its acclimation to this change (Benjamin et al., 1984).

Research Objectives

The aim of this study was to investigate the effects of substrate to inoculums ratio (SIR) on the AD of vegetable wastes using biochemical methane potential assays. The specific objectives were as follows:

1. To conduct a comprehensive literature review on the AD process and the importance on the difference factors affecting this process in particular with respect to SIR,
2. To set-up appropriate biochemical methane potentials (BMPs) to investigate the effect of difference SIR on the AD process during the three main phases,
3. To monitor the variations in total solids (TS), volatile solids (VS), volatile fatty acid (VFA), chemical oxygen demand (COD), soluble (sCOD) and on biogas generation and perform adequate statistical analysis to verify the level of difference on AD process dynamics for different SIR.

Pretreatment in AD Processes

The use of pretreatments is particularly useful in the digestion of biomass as these tend to have a high cellulose or lignin content. By manipulating the physical and chemical properties of these polymers, pretreatment method facilitates the breaking down of the substrates and aims at enhancing AD processes (Ward et al., 2008; Mudhoo, 2012).

Physical Pretreatment

The rate of hydrolysis and anaerobic biodegradability of substrates can be increased by the application of physical pretreatment. There are 3 major types of physical pretreatment;

1. Mechanical Pretreatment- it involves size reduction of the substrate which leads to the rupturing the cell walls, making the biodegradable components more accessible to microorganisms (Mudhoo, 2012). After the mechanical disruption, the intracellular components of the cells are immediately available for biological degradation, thus accelerating the AD process (Muller et al., 1998).
2. Thermal/ Thermochemical Pretreatment- it is a stage that improves the efficiency of AD process as thermal hydrolysis leads to the partial solubilisation of substrates, including the organic and inorganic compounds. During the thermal treatment, the digester volume is reduced and the biogas production is increased (Mudhoo, 2012). Thermochemical pretreatments use a combination of heat and chemicals to decrease particle size and improve solubilisation (Ward et al., 2008).
3. Ultrasonic Pretreatment- This technique is mainly used to break down complex polymers in the treatment of the different substrates used for AD (Warsd et al., 2008). It helps in increasing the methane production and reduces the volatile solids (VS) during the AD (Mudhoo, 2012).

Chemical Pretreatment

Chemical pretreatment enhances the AD treatment by hydrolysing the cell wall and membrane and thus increasing the solubility of the organic matter contained within the cells. Various chemical methods have been developed upon the different operating principles (Appels et al., 2008).

These include alkali pretreatments (with alkaline hydrolysis and alkaline peroxide), organosoly process, wet oxidation (treatment with water and air/oxygen at a temperature above 120°C for a period of time), ozonolysis pretreatment (treatment with ozone) and acid hydrolysis pretreatment (sulphuric acid is mostly used). In AD process, the alkaline pretreatment is mostly used as an increment at performer in terms of solubilisation of COD and biogas production (Mudhoo, 2012).

Anaerobic Digester Design

Several designs have been put forward for in-vessel AD of solid wastes (Cecchi et al., 1991; IEA, 1994). These technologies are classified in terms of factors such as solids concentration, mixing strategy, temperature and number of stages. These designs are illustrated below.

Single-Stage and Multistage System

In a single stage system, only 1 reactor is utilized for the microbial digestion of the organic matter, where all the three steps of biochemical transformations, i.e. depolymerisation, acidogenesis/acetogenesis and methanogeneis occur simultaneously (Gerardi, 2003). The multi-stage AD processes was introduced so as to improve digestion by having separate reactors for different stages of AD. This provides a better efficiency for each reaction that is the hydrolysis and the methanogenesis. Two reactors are normally used for these two reactions.

Batch System

In a batch digester, the digester is loaded with the organic material in one go, sealed, and allowed to degrade for a specific retention time which is related to temperature (mesophilic or thermophilic) and other factors. Once digestion is complete, the digestate is removed and the process is restarted (Erickson et al., 2004). In such system, the digester is allowed to go through all the degradation steps in dry modes, that is 30-40% TS (Monnet, 2003). The biogas that is evolved from such system, however, is not steady. There are three types of batch systems namely single stage batch system, sequential batch system and an Upflow Anaerobic Sludge Blanket reactor (UASB) (Monnet, 2003).

Process Parameters & Conditions Affecting AD

The metabolism of any process is affected by the process parameters. AD is a sensitive process in which all the process parameters should be in equilibrium among themselves so as to ensure a better efficiency of the digester and to have a maximum yield of biogas production by the proper degradation of the organic matter. The main parameters that affect the AD are discussed with their respective optimal ranges.

Particle Size

Particle size can influence the rate of AD by affecting the availability of a substrate (i.e. the surface area) to hydrolyzing enzymes, and this in particularly with plant fibres. Fibre degradation and methane yield improve with a decreasing particle size from 100 mm to 2 mm (Wards et al., 2008; Mshandete et al., 2006). Maceration of manure to decrease the size of recalcitrant fibres was found to give a better yield of biogas potential (16%) with a fibre size of 2 mm, and a 20% rise in biogas potential was observed with a fibre size of 0.35 mm. However, no difference was observed when the size of the fibres ranged between 5–20 mm (Angelidaki et al., 2000).

Total Solids (TS) Content

Total solids (TS) content is the amount of a fixed solid of a given sample. There are three ranges of TS content categorized in Table 1:

Table 1. Classification of AD systems according to TS content (Monnet, 2003)

% Solids	< 10%	15 – 20 %	22 – 40 %
Type of System	Low solids (wet) AD system	Medium solids (semi-dry) AD System	High solids (dry) AD System

Low solid digester systems have higher water requirements than high solid AD systems. It is more appropriate to start a set up with a low solid AD systems so as to allow the microbes to adapt themselves to their environment.

Hydrogen-Ion (pH) Concentration

The pH value gives information about a medium (substrate + inoculums) since it varies upon the buffering capacity of the medium itself (Sakar et al., 2009). pH plays a major role in anaerobic biodegradation in which pH influences the activity of microorganisms. Any fluctuations in pH such as an increase in acidity will reduce microbial growth and thus hinders the entire process (Gerardi, 2003). The optimum pH range for anaerobic digestion lies between 6.4 and 7.2 (Chugh et al., 1999). During digestion, the two processes of acidification (pH of 7 to 8.5) and methanogenesis require a lower pH levels for optimal process control (Rapport et al., 2008). The retention time of digestate affects the pH value. Since acetogenesis occurs rapidly in batch reactor, an accumulation of a large amount of organic acids can be observed, thus resulting in a pH below 5 which can lead to inhibition of the methanogenic anaerobes. However the carbonates ions forming during the reaction of CO_2 with water produces hydroxide ions and tends to buffer the process (Igoni et al., 2007). On the other hand, if the natural buffering system fails, manual addition of alkaline solution (potassium hydroxide KOH, and sodium hydroxide NaOH) is recommended to avoid the inhibition.

Temperature

The rate of decomposition and gas production is sensitive to temperature and generally, the process becomes more rapid at high temperatures (Igoni et al., 2007). It is important to maintain a stable operating temperature in the digester since sharp or frequent fluctuations in temperature affect the bacteria, mainly the methanogens. Process failure can occur at temperature changes in excess of 1°C/day and changes in temperature of more than 0.6°C/day should be avoided (Appels et al., 2008). Bacteria may be classified as psychrophic, mesophilic, thermophilic and hyperthermophilic depending on the temperature in which they function best (Nguyen, 2008). There are two most important temperature ranges in which the methane forming bacteria are activated. These are the mesophilic range, from 30 to 35°C and the thermophilic range, from 50 to 60°C. A change from mesophilic to thermophilic temperatures (or vice versa) can result in a quick decrease in biogas production until the necessary populations have increased in number (Ward et al., 2008). The operation in the mesophilic range is more stable and requires a smaller amount of energy expense (Khalid et al., 2011; Ward *et al.*, 2008). According to Gerardi (2003), the methane forming bacteria are inhibited at the temperature between 40 and 50°C. Temperature affects biological activity and this effect is mostly due to the impact of temperature on enzymatic activity or reactions.

Therefore an increase in temperature results in more enzymatic activity whereas a decrease in temperature results in less enzymatic activity (Khalid et al., 2011).

Carbon to Nitrogen (C/N) Ratio

Carbon to nitrogen (C/N) ratio is represented by the relationship between the amount of carbon and nitrogen present in organic materials. Microorganisms need nitrogen for the production of new cells. For methanisation, a nutrient ratio of the elements C:N:P:S at 600:15:5:3 is sufficient. The optimum C/N ratios in anaerobic digesters is between 20–30 so as to ensure sufficient nitrogen supply for cell production and the degradation of the carbon present in the wastes (Fricke et al., 2007; Sadi, 2010). For a high C/N ratio, the waste is said to be deficient in nitrogen which is an important nutrient for the growth of bacteria and it will thus affect the yield of biogas. A low C/N ratio indicates accumulation of excess nitrogen that is ammonia, resulting in an increase of the alkalinity level and thus disturbing the system (Sadi, 2010).

Volatile Fatty Acids (VFA)

Volatile fatty acids (VFA) represent a critical parameter in the operation and control of AD (Forster-Carneiro et al., 2008). It also gives an indication of the acid formed during acidogenic process (Madsen et al., 2011). However, the production of VFA can be toxic to micro-organisms, particularly to methanogens at a concentration of 6.7–9.0mol/m3. Their rising in concentrations are the result of accumulation due to process imbalances which can be caused by variation in temperature, organic overloading, toxic compounds, etc. (Appels et al., 2008). Moreover, the high biodegradability of the food and vegetable wastes (FVW) promotes a quick production of VFAs, resulting in a rapid decrease in pH, which in turn could inhibit the methanogenic activity (Garcia-Pena et al., 2011). It was suggested by Vavilin et al. (2003) that diffusion and advection of VFAs inhibited both polymer hydrolysis and methanogenesis thus increasing the initial hydrolysis rate above a critical value which causes an inhibition to methanogenesis and then hydrolysis.

Volatile Solid (VS)

Volatile Solids (VS) is an important factor that must be known before considering any AD process. VS measure the biodegradability of the substrate which is to be digested. The percentage of VS depends on the different composition of wastes that affects the biogas yield and quality, as well as the digestate quality (Lahdheb et al., 2009). During the digestion process, volatile solids are degraded to a certain extent to produce biogas (Appels et al., 2008). High VS are best suited to AD treatment (Binod, 2008).

Moisture

High moisture contents usually smooth the progress of the AD. However, it is difficult to keep the same availability of water throughout the digestion cycle (Hernandez-Berriel et al., 2008). As the process of AD proceeds, the initial high moisture content decreases to a lower level.

The high water content affects the process since the readily degradable organic matter will get dissolved (Bouallagui et al., 2003). However, a strong leachate is found in bioreactors under the 70% moisture regime thus producing a high rate of methane (Khalid et al., 2011).

Retention Time

The microorganisms require a certain amount of time to digest the organic matter and to reach a desire level of treatment (Gerardi, 2003). The longer a substrate is kept under proper reaction conditions, the more complete its degradation will be. However, the rate of the reaction will decrease with increasing residence time, showing that there is an optimal time for the achievement of the digestion benefits in a cost effective way. This appropriate time depends on the feedstock, environmental conditions and intended use of the digestate (Haleh, 2011).

Roles of Microorganisms

The AD process can be catalyzed by microorganisms that convert complex macromolecules into low molecular weight compounds. Inoculum is crucial for the optimization of the waste/inoculum ratio. Sludge is a common inoculum used for the treatment of waste although naturally selected strains or artificially mixed strains of microorganisms are also used. Additionally, cell aggregates in the form of flocs, biofilms, granules and mats, with typical dimensions range of 0.1 to 100 mm may also be employed in the treatment system (Khalid et al., 2011). According to Fricke et al. (2007), organic material is most likely to decompose by heterotrophic microorganisms and it is reported by Lee *et al.* (2009) that Clostridium species are the most common type of degraders under anaerobic condition.

However, biological treatment rarely relies solely on a single microbial strain; microbial consortiums are generally responsible for AD process (Fantozzi et al., 2009). It has also been found that high concentration of organic acid like acetic acid (>5000 mg/L) and butyric acid (>3000 mg/L) in the biodigester slow down the growth of microorganisms resulting in low production of energy rich compounds (Kim et al., 2008).

Biogas

The major product from the process of AD of organic matter consists mostly of methane and other gases. The biogas can be used to supply heat, electricity, process steam and even in the manufacture of methanol.

Table 2. Biogas composition (Arthur et al., 2010)

Gases	Percentage composition by volume (%)
Methane (CH_4)	40-75
Carbon dioxide (CO_2)	25-40
Nitrogen (N_2)	0.5-2.5
Oxygen (O_2)	0.1-1
Hydrogen sulphide (H_2S)	0.1-0.5
Ammonia (NH_3)	0.1-0.5
Carbon monoxide (CO)	0-0.1
Hydrogen (H_2)	1-3

The biogas burns with a blue flame and when it has a methane content ranging from 60% to 70%, it is said to have a heating value ranging between 22,000 kJ/m^3 and 26,000 kJ/m^3 (Brown, 2001). Biogas yield from biodigestion depends on the substrate composition, type of substrate, retention time and biodigester conditions (Arthur et al., 2010). The average composition of biogas is shown in Table 2.

Pure Methane gas is colorless and odorless. The odor which is evolved when biogas is produced is due to the hydrogen sulphide (H_2S) content. If high quantity of H_2S present in biogas, it is recommended to remove the gas as it is corrosive in nature. Also, when removed, it increases the energy content of the biogas. Methane gas is used mainly as a stationary fuel rather than a mobile fuel as it is very difficult to compress (Adawi, 2008).

Digestate

Digestate is the solid residues of the original input material to the digesters that the microbes cannot use. Its main advantage is that it has a high nutrient content. Its quality should be acceptable for purposes like soil enrichment or landscaping. Digestate quality can be categorized into 3 criteria: chemical, biological and physical aspects (Monnet, 2003). The chemical aspects of the quality management of digestate are basically related to the presence of:

1. Heavy metals and other inorganic contaminants,
2. Persistent organic contaminants,
3. Nutrient (NPK)

Toxicity Associated with AD

Some toxic gases associated with AD are discussed below.

Ammonia

The production of ammonia is due to the degradation of the nitrogenous matter which is mostly in the form of proteins and urea (Kayhanian, 1999). The two major forms of inorganic ammonia nitrogen in aqueous solution are the ammonium ion (NH_4^+) and free ammonia (NH_3). NH_3 has been reported to be the main cause of inhibition as it is freely membrane-permeable (de Baere et al., 1984).

Sulphide

Sulphate is a common component within the AD industries (O'Flaherty et al., 1998). Sulfate is reduced to sulphide by the sulphate reducing bacteria (SRB) in the reactor. There are two major group of SRB; incomplete oxidizers which reduce compounds such as lactate to acetate and CO_2 and the second group is complete oxidizers which convert completely the acetate to CO_2 and HCO_3^- (Koster et al., 1986).

Light Metals Ions

The light metal ions such as sodium, potassium, calcium, and magnesium are present in the influent of anaerobic digesters. They are usually formed by the breakdown of organic matter (such as biomass), or pH adjustment with chemicals (Grady et al., 1999). The degree of toxicity of salt of the metals was found to be predominantly determined by the cations of the salt. However, they are important for the microbial growth and influence the growth rate just like other nutrients.

Heavy Metals

Chromium, iron, cobalt, copper, zinc, cadmium and nickel are identified as subjects of particular concern. Since they are not biodegradable, they can be accumulated to toxic concentrations. It was reported that heavy metals are often the main cause of digester failure. The heavy metals disrupt the function of enzymes and form structures by biding with thiol and other groups of proteins (Jin et al., 1998).

Co-Digestion

Since its development, the AD had been used for the digestion of a single substrate. But quite recently, new technological research and development have pushed forward their limit to embrace the concept of co-digestion. Co-digestion is the degradation of different mixture of two or more substrates in the same reactor (Jingura et al., 2007). Digesting different substrates simultaneously has got many advantages in terms of ecological, technological and economic aspects since such system is associated with better fertilizer quality with higher nutrients in the digestate and an increase in biogas potential yield hence increasing the economic viability of the process.

Effects of Substrate to Inoculum Ratio (SIR) on AD Systems

In the previous research on the effects of substrate to inoculums ratio was limited. Some of the biodegradability tests used for the determination of anaerobic biodegradability of micro-pollutants and complex substrates have been introduced after 2002. Litterature showed that research on the effects of substrate to inoculums ratio was limited. Some of the biodegradability tests used for the determination of anaerobic biodegradability of micro-pollutants and complex substrates have been introduced after 2002 (Muller W.R., 2004). According to Fernandez et al. (2001) in a start up phase, the greatest values of the specific methane production were found with lowest values of the SIR. Most of the ISRs used in the previous studies were 3.0, 2.0, 1.5, 1.0, 0.8 and 0.5 (Raposo et al., 2009). Several types of inoculums (based on various studies done around the world) are used in the AD process so as to maximize the biogas production from different mixtures of substrate and inoculums. From a study done by Poggi-Varaldo et al. (1997), three types of inoculums were used: cattle manure, soil and waste activated sludge. Neves et al. (2004) used ISR range of 2, 1, 0.74 and 0.43 with two types of sludge inoculum (granular and suspended). The granular inoculum prevented acidification of the reactor over the range of SIR used, whereas the suspended sludge avoided the acidification at higher level of SIR. Digester sludge from a municipal wastewater treatment plant may also be used (Raposo et al., 2009).

Therefore, the aim of the current research is to look at the SIR as the one of the most important factors likely to influence the result of a BMP test carried out using batch array and to monitor the progress of the anaerobic conversion by standardized procedures. In this case, vegetable waste is chosen as the substrate since it is produced in large amount during market days and vegetable waste has a high biodegradability since it is composed of organic matter.

Materials and Experimental Methods

The purpose of this study was to analyze the effects of inoculums to substrate ratio on vegetable wastes in an anaerobic state. In order to achieve the given aim, 4 anaerobic reactors were set up for comparison as categorized in Table 3.

Table 3. Showing the different composition of the mixture

Reactors	Substrate (%)	Inoculum (%)
A	100%	0%
B	75%	25%
C	50%	40%
D	0%	100%

For this study, operational and performance parameters in anaerobic digestion were monitored, leading to a systemic analysis of VFA, pH, TS, VS, COD, sCOD, biogas production as well as gas composition (Figure 1). Tests of TS, VS, VFA, pH, COD and sCOD were carried out twice a week. The biogas production was monitored everyday till the gas evolvement was stabilised.

Materials for the Anaerobic Reactor

2.0-Liter plastic bottles made air tight by the use of a thick rubber septum, were used as anaerobic reactor for the BMPs. Dark thick plastic sheets were wrapped round the reactor so as to prevent the penetration of light. Red silicon gasket making was used to seal the cork, preventing any leakage of the BMPs and penetration of air in the reactors. To enable sampling from the BMPs, plastic tubing was fixed through a hole drilled in the cork.

Gas Collection

Gas collection was carried out by the water displacement method as shown in Figure 2. The BMPs generated from the reactor, travels through the plastic tubing and rises in the inverted cylinder. The measuring cylinder was filled to the top with water before carefully inverting it into a plastic bucket containing around ¼ level of water. Atmospheric pressure pushing down on the water in the plastic bucket keeps the water in the inverted cylinder from coming out. As the gas rises in the cylinder, its pressure pushes the water downwards. The gas volume was directly recorded from the graduated measuring cylinder used.

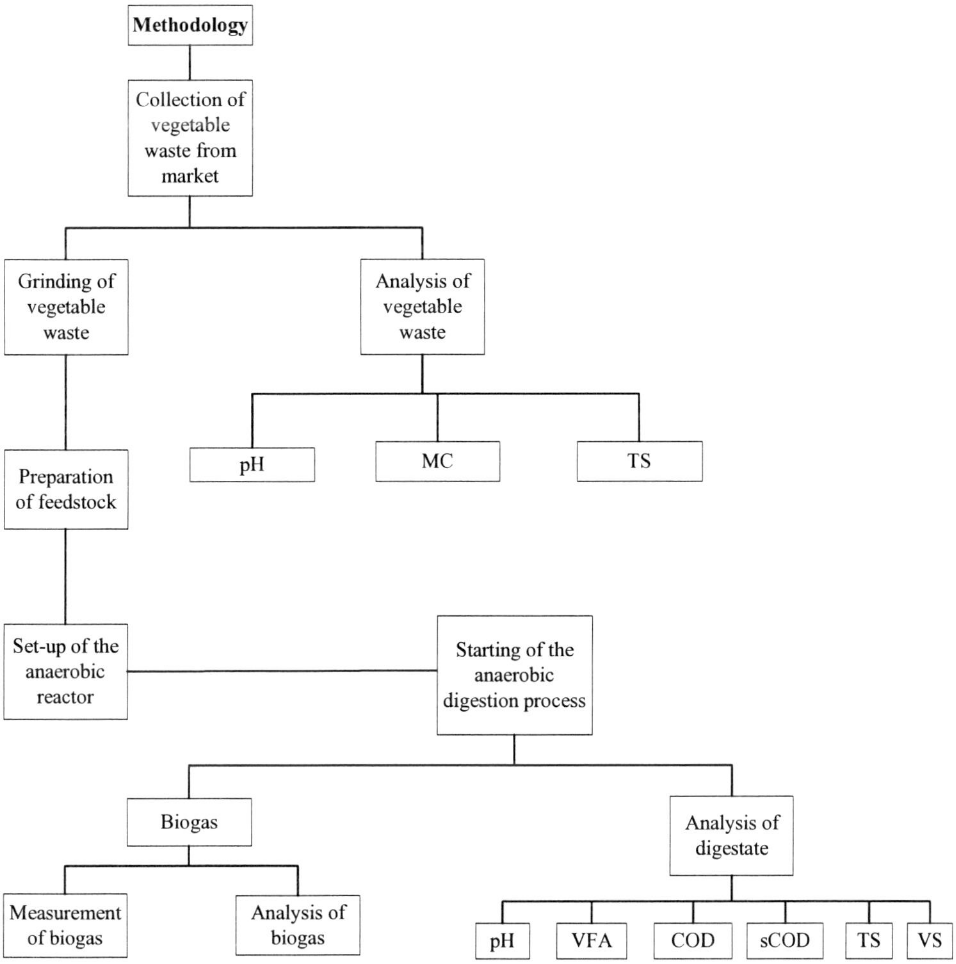

Figure 1. Methodology for the feed preparation and the digester set-up.

Preparation of the Substrate

Vegetables wastes were collected at the market of Chemin-Grenier in the South of Mauritius. The moisture content of mixed vegetable wastes (carrots, cabbages, cauliflower's leaves, beetroot, tomatoes and potatoes) was determined so as to properly adjust the amount of water and innoculum to produce a better mixture with the appropriate solids and moisture content.

The different SIR used was based on previous studies carried out in the field of anaerobic digestion. A Final TS of 8% was found apt according to literature values. As a means of enhancing the digestion process, the substrates that were to be placed in the batch reactor were first reduced in size.

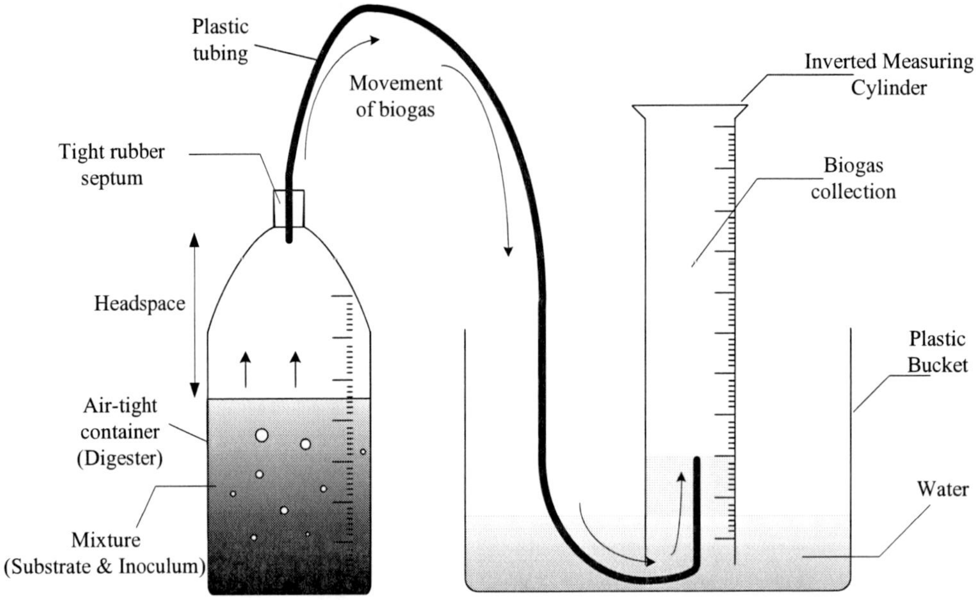

Figure 2. Experimental Set-up for the collection measurement system of biogas.

Size reduction may be achieved by the adoption of either mechanical technique like shredder or grinder, or manually through the use knives. In this study, grinding was preferred for it yielded a more uniform size. The prepared mixture was then placed in the 2L reactor with an allowance 0.5L to account for the accumulation of the biogas.

Analytical Techniques

Moisture Content and Total Solids

Determination of the percentage of MC and TS was carried out in the following way:

- Raw samples were taken from the reactors and a known mass of the sample was put in the oven and allowed to dry to a constant mass at 105°C for 24 hours.
- After the drying, the difference in weight before and after was calculated and %MC and %TS were computed as follows:

$$\text{Moisture Content, \% M.C} = \left(\frac{\text{Loss in weight after drying}}{\text{Net wet weight}}\right) \times 100$$

$$\text{Total dry solids, \%TS} = 100\% - \text{\%M.C}$$

- ***Volatile Solids***
- The dried sample obtained from the Oven was weighed accurately and put in a crucible. The weighed sample was then ignited at 550°C in the furnace for a minimum of 2 hours so as to ensure complete burning. After ignition, the sample was

weighed and the volatile solids content were calculated by the difference in the weight before and after ignition.

$$\text{Volatile solids, \%VS} = \left(\frac{\text{loss in weight after ignition}}{\text{net dry weight}}\right) \times 100$$

$$\text{Ash content, \%Ash} = 100\% - \%\text{VS}$$

Volatile Fatty Acids, (VFAs)

10 ml of raw sample from the reactor was placed in a flask and diluted up to 60 ml with distilled water. The diluted sample was then stirred on a magnetic stirrer so as to have a uniform solution. With the help of a pH-meter, this uniform solution was titrated with acids (HCL) and subjected to five pH titration points. The pH values and the titrant values were plugged into the titgram program by which the VFA content and the alkalinity were calculated.

Chemical Oxygen Demand (COD)

The sample taken from the reactor was diluted to 100 or 200 times due to its high value of COD. Potassium dichromate VI reagents followed by the catalyst reagent was added to 2 mL of diluted sample in the COD test tubes. It was then heated for 2 hours in the COD block digester. After heating, the resultant sample was allowed to cool before its titration against Ferrous Ammonium Sulphate (FAS) solution.

Soluble Chemical Oxygen Demand (sCOD)

Raw samples were taken from the reactor and were centrifuged for around 30 minutes at 2000 rpm (model: R-BC laboratory centrifuge). After the centrifugation, the samples were filtered and the filtrates obtained were diluted to the required dilution factor (100 to 200 times). 2 mL of the diluted filtrate was tested for the sCOD following the same procedures as the COD test.

pH

Raw samples taken from the digester were diluted with distilled water and the pH was measured directly by making use of a pH meter.

RESULTS AND DISCUSSIONS

The physical and chemical characteristics of the organic waste are important for the design and operation of anaerobic digesters as they affect the biogas production and the stability of the anaerobic digestion process. In this section, the experimental results are analyzed and discussed. pH, VFA, COD, sCOD, VS, TS are the parameters that were experimentally analyzed for 34 days whereas the gas production was monitored daily for 60 days. 8 reactors were set up; with 4 of them denoted for sampling to perform tests and 4 identical ones monitored for the biogas production daily (Table 4).

Table 4. Composition of reactors for amount of inoculum added (by volume)

Set Up	Composition
A	0% Inoculum
B	50% Inoculum
C	75% Inoculum
D	100% Inoculum

Mechanical Pretreatment

The vegetable waste was ground in order to reduce its size and to have a homogeneous mixture before being mixed with the inoculums. Upon size reduction, a larger surface area of the wastes is offered thereby enhancing the degradation rate. Furthermore, mechanical treatment makes use of several strategies for physically disintegrating the cells and partly solubilising their content (Appels *et al.*, 2008). In so doing, the intracellular of the cells are available for the biological degradation and as a result accelerate the process of digestion (Muller *et al.*, 1998).

Properties of Inoculum

From literature, it was found that physical and chemical characteristics of the organic wastes undergoing anaerobic digestion are of major importance for the design and operation of the reactors, mainly on the biogas production and stability during anaerobic digestion (Garcia- Pena *et al.*, 2011). An initial characterization of the inoculum (cow dung) was carried out and the results presented in Table 5. While Wang *et al.* (2011) reported a range of 9.7-13.0% for total solid contents of cow dung; the total solid content of the inoculum used in this study was characterized by a value of 8.0%, which falls slightly below the lower limit of the range and corresponding moisture content of 92.0%.

Table 5. Initial characteristics of inoculum

Moisture content (%)	92.0
Total Solid (%)	8.0
VS (%)	50.0
COD (mg/L)	18000.0
sCOD (mg/L)	4316.55
pH	9.05

This may be explained by the fact that characteristics of inoculum varies according to the source of inoculum and the methods, processes and conditions through which the inoculum are being handled and stored (Wang *et al.*, 2011). Dennis (2001) also reported that the value of VS for cow manure to be 83.0% and the value obtained during the experiment was quit low, being at 50.0%. This low value is due to the high moisture content.

Properties of Substrate

Grinded substrate was also initially characterised. Compared with the inoculum, the grinded substrate showed a lower solid content owing to a higher moisture content and lower organic fraction than that of the inoculum (Hawkes *et al.*, 1980), as well as lower COD and sCOD values as compared to the innoculum. Upon intial pH analysis, the innoculum used exhibited alkaline characteristics (pH 9.05) while the vegetable used was slightly acidic (pH 6.0). Such acidity may be attributed to the presence of tomatoes which, according to Kendell (2006), are considered to be a high-acid food of pH below 4.6. André *et al.* (2007) also mentioned that the degree of aciditiy is directly proportional to the amount of tomatoes used in a mixture, implying an increase in acidity when higher amount of tomatoes are utilized. Furthermore, the inoculum got a higher COD and sCOD content than the grinded substrate. Also, the initial of the inoculum is alkaline (pH 9.05) whereas that of the vegetable waste is slightly acidic (pH 6.0). This acidity may be due to some vegetables that were present in the waste such as tomatoes. It is reported by Kendall, 2006 that tomatoes are considered as a high-acid food (pH below 4.6). Tomatoes are also considered as a rich source of lycopene and vitamins (Khan *et al.*, 2006). Furthermore, if there are more tomatoes in a mixture, the acidity of that mixture will be increased. Therefore the degree of acidity is directly proportional to the amount of tomatoes in a mixture (André *et al.*, 2007).

Table 6. Initial characteristics of substrates

Substrate	
Moisture content (%)	93.60
Total Solid (%)	6.40
VS (%)	76.92
COD (mg/L)	11870.50
sCOD (mg/L)	3600.0
pH	6.00

pH Variation

The value and stability of pH in an anaerobic digester is very important; it indicates the process stability and proper activities of methanogens. According to Bouallagi *et al.* (2010) and Van Haandel *et al.* (1994), methanogenesis only proceeds at a high rate when pH is maintained within the neutral or alkaline range. Figure 3 describes the pH variation that occurred within this study for a period of 30 days. An initial pH drop was observed for all the setups and this may be explained by the hydrolysis and conversion of the easily digestible fraction of the organic matter at the start of the reaction (Macias-Corral *et al.*, 2008). Generally, each of the different group of micro-organisms has a different optimum pH range.

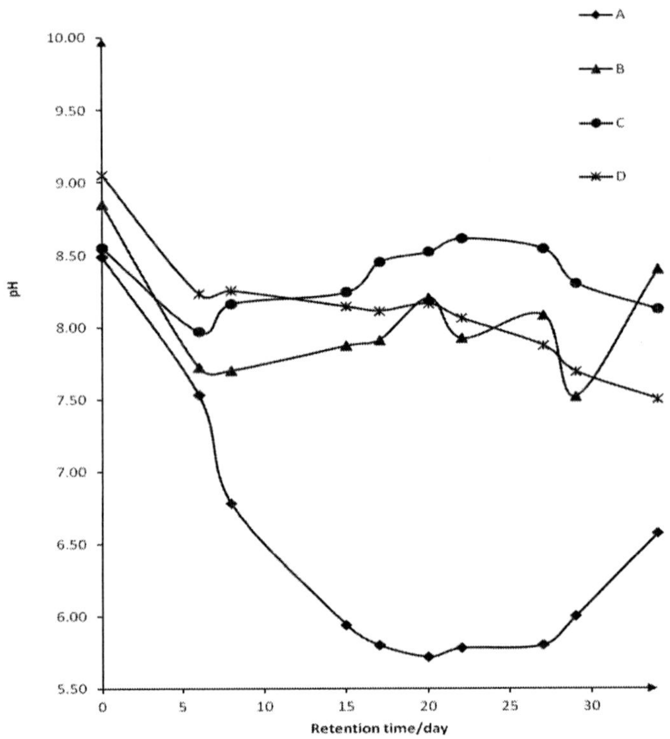

Figure 3. Variation of pH with time.

Methanogenic bacteria are very sensitive to pH with an optimum values lying between 6.5 and 7.2 (Turouskiy, 2006). According to Hwang *et al.* (2004), the fermentative microorganisms are to some extent less sensitive and can function in a more extensive range of pH between 4.0 and 8.5. Therefore at a low pH the main products are acetic acid and butyric acid, while at a pH of 8.0 mainly acetic acid and propionic acid are produced (Boe, 2006). The pH in the effluent varied between 8.0 and 9.0 indicating the process stability and the optimal activity of methanogenic bacteria. This result showed that the digestion mixture systems were well buffered. After the initial drop, the pH began to rise gradually. This is because the fatty acids were transferred to the methane phase reactors and were consumed by the methanogens. The variation of the pH during the experiments was caused by the periodic accumulation of the fatty acids in the solid phase reactor and the subsequent transfer and consumption of VFA by methanogenesis (Macias-Corral *et al.*, 2008). If the pH had dropped to 4-5, digestion would have stopped completely (Brakel, 1980). B, C and D showed similar pH variations ranging from (8.55 and 9.05), while the setup containing wastes only (A) showed a much higher drop in pH (up to 8.49) before a rise was observed. The reason for such low pH values in the reactor might be because of the high VFA production and low buffering capacity (alkalinity) of the waste (Nguyen, 2006).

Chemical Oxygen Demand

The chemical oxygen demand (COD) was used to measure the degree of decomposition of organic matter that is present in the digester for an experimental period of 34 days. Generally, COD is a measure of the amount of soluble organic compounds that give an indication of the amount of soluble carbon compound which can be converted to methane throughout the process of anaerobic digestion (Schnürer et al., 2010). In all the four reactors, the COD values were very high with reactor A (0% Inoculum), B (50% Inoculum), C (75% Inoculum) and D (100% Inoculum), yielding COD values of 18000.00 mg/L, 17266.19 mg/L, 13000 mg/L and 11870.50 mg/L respectively. These COD values indicated that they were highly contaminated with organic matter as was equally observed by Sarkar et al. (2009). This contamination may be due to the vegetable waste used in this study which was still in a fresh state; The high amount of COD in all the samples indicated that there was a large amount of organic matter present in each sample. From the graph, it can be observed that the amount of COD gradually declined with time which may be explained by the fact that COD was being consumed by fermenting and methanogenic bacteria and converted into biogas (Macias-Corral et al., 2008).

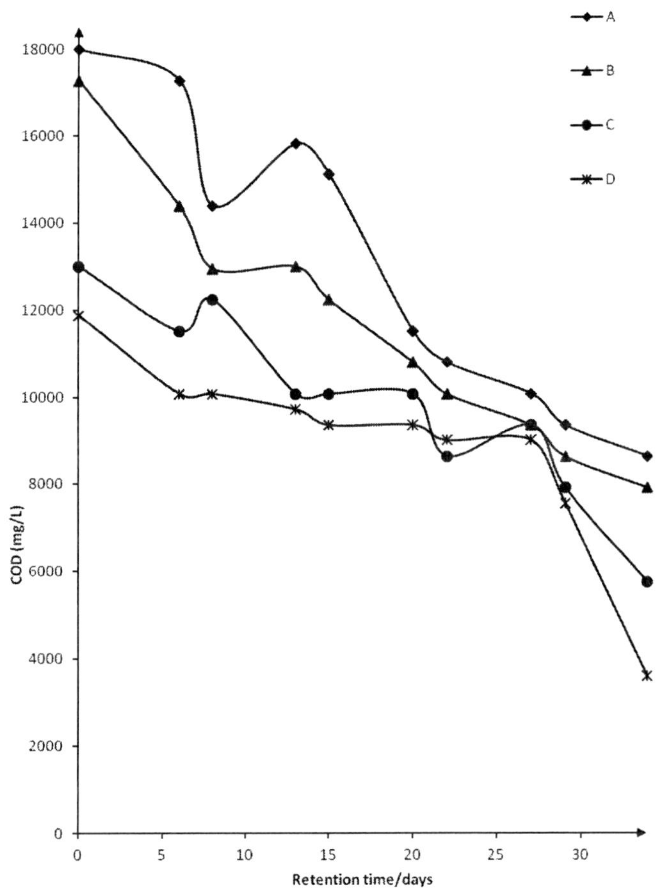

Figure 4. COD variation over 34 days.

Throughout the 34 days, this general decreasing trend of COD could be observed. There was an increase in COD from 14,388.49mg/L to 15,827.34mg/L that was observed from Day 8 to Day 13 for the sample containing 0% inoculum. This amount was soon decreased showing that these organic matter were digested by the bacteria to produce biogas. The increase in COD, from 12,949.64mg/L to 13,000.00mg/L from Day 8 to Day 13, for the second sample was also due to accumulation of organic matter but because of the 50% inoculum present, these organic matter were readily digested. For the last sample containing 100% inoculums, the COD was decreasing constantly. Due to the presence of high amount of microorganisms in the inoculums, the readily available organic matter was converted to biogas rapidly. For the 100% inoculum, the initial COD was 11,870.50 mg/L with a decrease observed as shown by the analysis. A decrease of around 15% has been noted from day 0 to day 8. It is also observed that from day 27 till day 34, there is a sharp decrease in the COD. This might be because of the endogenous bacterial activity among the micro-organisms in the reactors, feeding on each other (Jackson et al., 2003).

Soluble Chemical Oxygen Demand

On day 0 (set up day), the 0% Inoculum started with a soluble chemical oxygen demand (sCOD) of 4,316.55 mg/L.

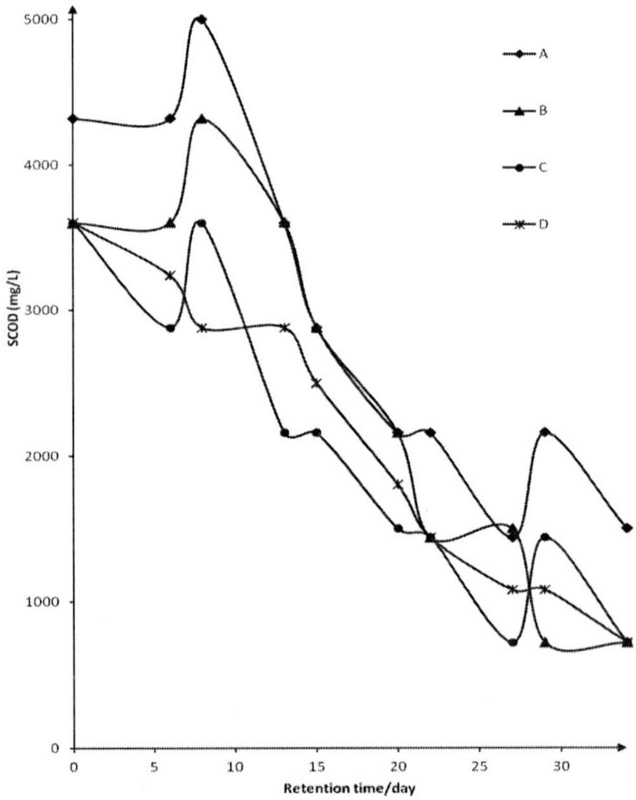

Figure 5. sCOD variation over 34 days.

A maximum value was attained on day 8 before the sCOD of the reactor's content started to show a decreasing pattern in the end of the experiment, with some fluctuations between day 22 and 34. On day 8, degradation of the vegetable wastes was highest indicating a high fraction of biodegradable organic matter. sCOD concentrations increased due to the production of organic acid during the fermentation process. These acids formed are assumed to catalyze further the hydrolysis of the hemicelluloses (Ozkan et al., 2011). The maximum sCOD removal was obtained from the digester that operated with 100% inoculum. However, the degrading of sCOD for all four set-ups has a decreasing order.From the graph, it was observed that the reactors which contained the mixture of inoculum and substrate along with that of the 0% inoculum reached their maximum sCOD on day 8 before a gradual decrease started. For the 100% inoculum, the sCOD decreased gradually till day 34. This observation might be due to the degradation within the inoculums itself. It was associated with the least volume of biogas produced as the degradation rate of the inoculum was low. It was also observed that on day 34, three of the reactors (50% inoculum, 75% inoculum and 100% inoculum) had the same sCOD value, 719.42 mg.

Ratio of Soluble COD to Total COD

Comparison of soluble COD to Total COD (SCOD/TCOD) ratios gives a general indication of the extent of hydrolysis (Park *et al.,* 2010). In this study, the SCOD to TCOD ratios showed an increasing trend for the first 8 days and between day 8 and day 34, the ratios shifted to a decreasing and fluctuating trend. The increase in SCOD/TCOD ratio was highest for reactor D, followed by reactor C, A and finally B. Reactor D, which was composed of 100% inoculums, had the highest intial SCOD/TCOD ratio. Also, with C having a higher % of innoculum (75%) than B (50%) and exhibiting a higher sCOD/TCOD ratio, a probable directly proportional relationship between sCOD/TCOD ratio and % of inoculation may be deduced.

Volatile Fatty Acids

In the AD processes, volatile fatty acid (VFA) are the most important intermediates which are degraded by proton-reducing acetogens with the involvement of hydrogen consuming methanogenic bacteria (Mechichi et al., 2005). volatile acids represent an important parameter in the operation and control of an anaerobic reactor (Forster-Carneiro et al., 2008) and the total volatile fatty acid (TVFA) concentration has been found to give a very good indication of the metabolic status of an anaerobic degradation process (Fernandez et al., 2005; Bjornsson et al., 2000). Throughout the digestion process, the levels of TVFA at steady-state were very low. These values show a high stability in the reactors after an initial transitory increase in TVFA.

The production of large amounts of VFA was due to the rapid acidification of vegetable waste which resulted to the inhibition of the biological activity of the methanogens (Bouallagui *et al.*, 2005). This occurs typically for the start-up of an anaerobic process when the balance of the hydrolytic bacteria, fermentative bacteria and methanogens has not yet stabilised (Hartmann and Ahring, 2005).

The presence of a stable pH and alkalinity of the digestors confirmed this stability. It has been reported by Brakel (1980) that high concentrations of VFA are inhibitory to the methane bacteria.

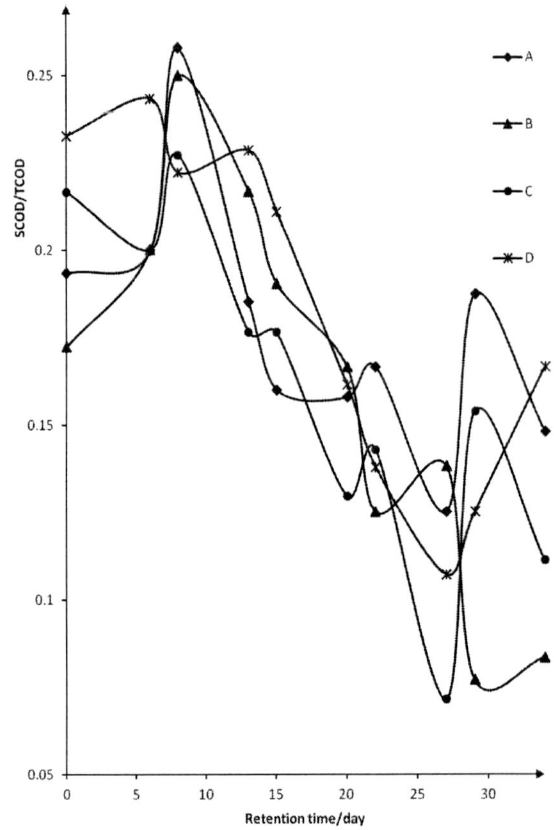

Figure 6. SCOD/TCOD variation over 34 days.

Figure 7 shows the variation of VFA with time during 34 days. From the beginning of the monitoring till day 8, the VFA had an increasing trend for reactors A, B and C. This was probably because of the accumulation of organic acid formed by the substrate (André et al., 2007). These acids are mostly considered to catalyse the hydrolysis of the hemicelluloses found in the vegetable waste (Ozkan et al., 2011). Reactor D had an almost constant VFA since it consisted of inoculum only. Reactor A had the highest VFA followed by B, C and finally D. It could be deduced that the VFA produced is directly proportional to the proportion of vegetable wastes present within the reactor given that the presence of organic acids has been linked with the degradation of the organic wastes (André et al., 2007).

Volatile Solids

Volatile solids (VS) monitoring is an important control for the measurement of the biodegradation of organic substrate, which indirectly shows the metabolic status of the microbial groups in the anaerobic reactor (Elango et al., 2007).

The system performance and the removal efficiency of the organic materials during the process is a measure of the amount of solid destroyed during anaerobic digestion.

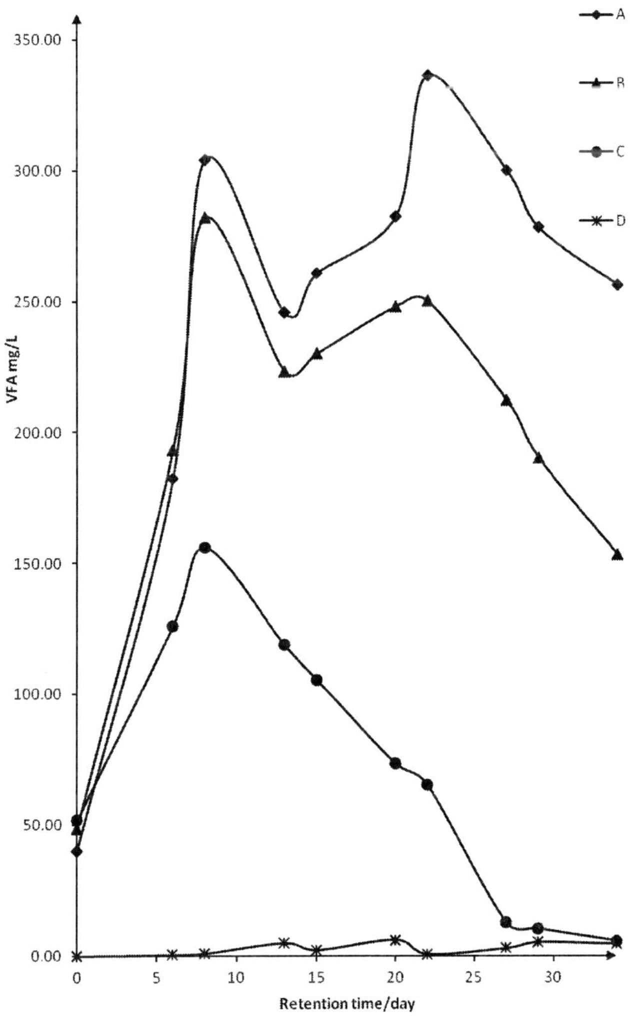

Figure 7. VFA variations with time of digestion.

The vegetable wastes showed excellent digestibility and seemed to accelerate the digestion process. A highest VS degradation efficiency of 77.78% was obtained for the vegetable wastes inoculated at 50% cow dung and it was associated with the highest overall specific biogas production for the period of 60 days as explained later on. It is very likely that the high degradation efficiency in the mixture (substrates and inoculums) was due to an improved ratio of nutrients and better availability of the organic substances; this facilitates their assimilation by anaerobic anaerobes and increases the degree of degradation (Krupp et al, 2005).

A VS degradation efficiency of around 71% was obtained a ratio of 75% substrate and 25% inoculum. On the other hand, a lowest efficiency of 48.3% was obtained by the 100% inoculated setup. This high efficiency is also associated with the highest amount of biogas produced while the lowest efficiency is associated with the minimal biogas produced in that period of time (34 days).

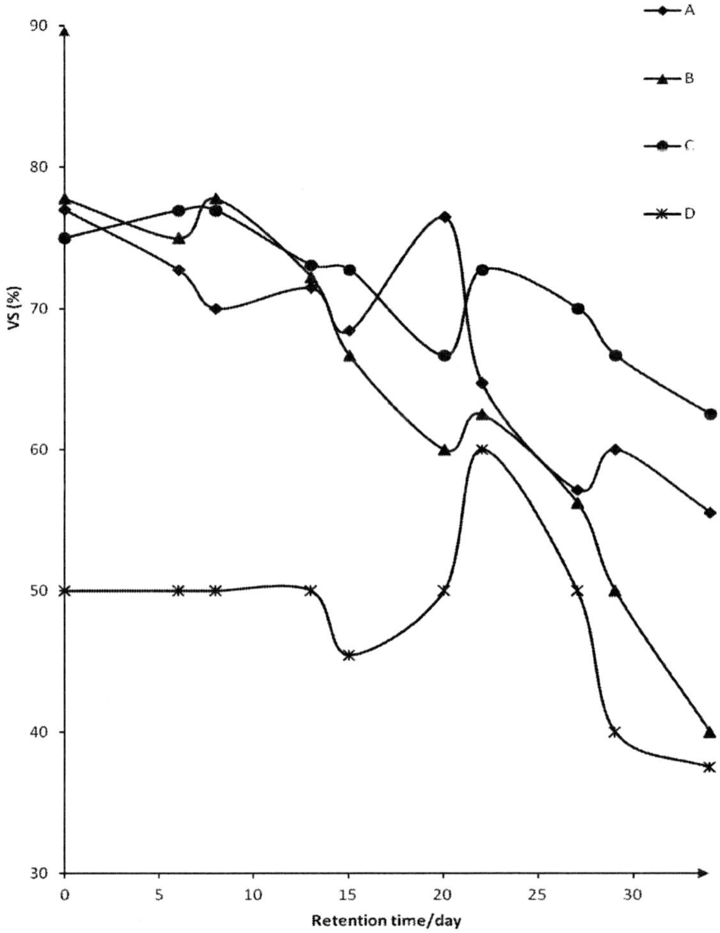

Figure 8. Variation of volatile solids in time.

Cumulative Biogas Production

The biogas evolution was monitored daily. This was done so as to prevent any explosion during the accumulation of biogas in the reactor. Figure 9 shows the biogas evolvement during the monitoring days.

The biodegradation of the organic waste was notified since the first day of setting up of the reactors. During the start-up, it was observed that the biogas production was low and increased gradually.

The gas production rate from reactor A (0% inoculum) was higher than the other reactors. In the 100% inoculum reactor sodium bicarbonate was not used to increase the pH as compared to the 100% substrate which had a low acidity. Such action might have led to the release of CO_2 from reactions of sodium bicarbonate within the reactor.This tallies with the concentration of CO_2 obtained (55%) when the gas from reactor D was collected and analysed. The possible explanation might be because of the early micro-aeration in pre-stage which might have resulted in better hydrolysis or acidification during start-up of methanization period providing substrate for methanogens (Nguyen P.H.L., 2006).

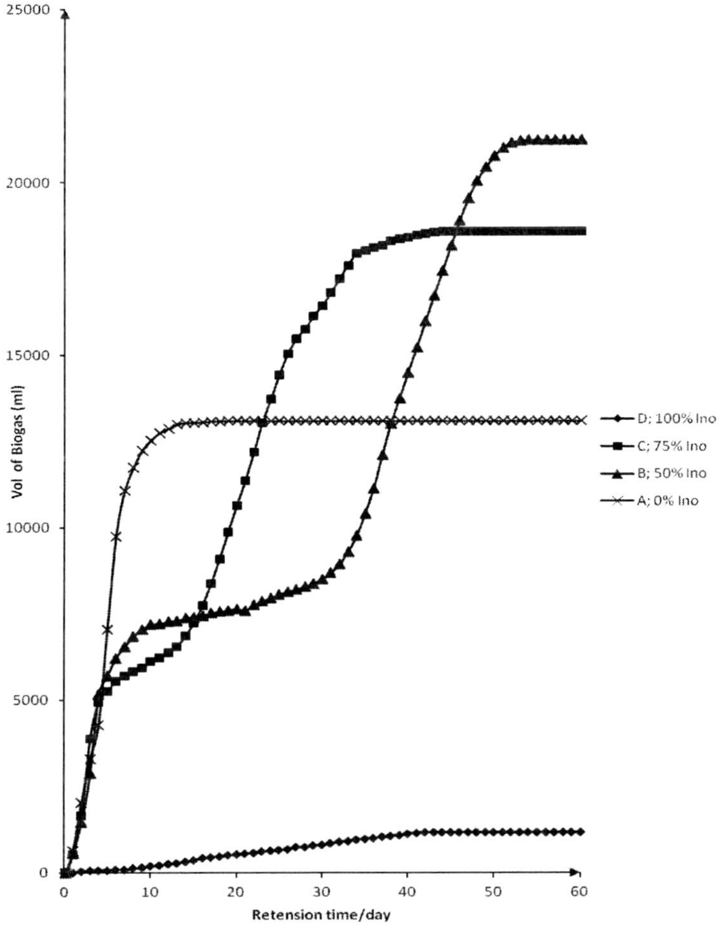

Figure 9. Cumulative biogas generation..

Composition of Biogas

According to Arthur *et* al. (2010), the percentage of methane lies between 40% and 75 %. The biogas evolved from the 50% and 75% Inoculum digester was collected using urine bag. But for reactor A, there is only 5% of CH_4 and 55% of CO_2. This is because of the sodium bicarbonate which was used for decreasing the acidity of the A and made it more alkaline since acidity would reduce microbial growth and thus hindered the entire process (Gerardi, 2003). This bag was kept for several days so that there was enough gas to be analysed. In digesters B and C, it can be seen that there was no H_2S evolved and only 1 ppm of CO while in that of A, 1ppm of H_2S was produced with the 75% Inoculum, a higher percentage of CH_4 was obtained. This is might because of the higher percentage of bacteria which would have degraded the 25% of vegetable waste more rapidly than that within the 50% Inoculum digester. An additional observation was the presence of oxygen despite the fact that the system should have been anaerobic. Such event may be attributed to the infiltration of oxygen in the digester during sampling.

Table 7. Biogas composition

Gas composition	A; 0% Ino/%	B; 50% Ino /%	C; 75% Ino/%
CH_4	5	46.6	70.9
CO_2	55	9.3	10.4
O_2	11	9	5.1
balance	25.8	35.9	13.5
H_2S	1ppm	0 ppm	0 ppm
CO	1ppm	1 pmm	1 ppm

CONCLUSION

Over the years, anaerobic digestion has received considerable attention as a means of waste treatment for stable bio solids and which is also considered as an effective tool for both waste reduction and energy production mainly for biogas. Past studies have explored the potential of various substrates including cattle manure, soil and waste activated sludge. Biochemical methane potential assays (BMPs) provide a practical estimation digestibility of a given substrate and this present study aims at estimating biomass potential of selected feedstock for biogas production from laboratory experiments. Adopting this technology on large scale can reduce the amount of organic matter being dumped in the landfill, around 80% in the case of Mauritius according to a study by Mohee (2002). Vegetable waste was used as substrate for the process of anaerobic digestion in view to obtain an estimate of the amount of biogas produced in case AD is adopted as a future waste treatment in Mauritius. The anaerobic digestion processes were also analyzed for the effects of substrate-to-inoculum ratio. The results from this study help to understand the importance of using a correct substrate-to-inoculum ratio in order to obtain a better control of the process. In this study, a 1:1 subsrate-to-inoculum ratio on volume basis gave the highest cumulative volume of biogas and the process dynamics proceeded consistently.

REFERENCES

Adawi O.A., 2008. Design, Building and Techno-Ecomonic Evaluation of Biogas Digester. Thesis (Msc) An-Najah National University Faculty of Graduate Studies. [online] Available from http://scholar.najah.edu/sites/scholar.najah.edu/files/all-thesis/5171585.pdf.

André, M.K.P., Márcia, M.C.F. *Analysis Chimica Acta*, 2007, 595, 221-227.

Angelidaki, I., Ahring, B.K. *Water Science and Technology,* 2000, 41, 189–194.
Appels, L., Baeyens, J., Dewil, R. *Progress in Energy and Combustion Science*, 2008, 34, 755-781.
Arthur, R., Baidoo, M.F., Antwi, E. *Renewable Energy,* 2010, 38, 1510-1516.
Benjamin, M.M., Woods, S.L., Ferguson, J.F. *Water Research*, 1984, 18, 601-607.
Binod, K.C., 2008. Dry continuous anaerobic digestion of municipal solid waste in thermophilic conditions, Asian Institute of Technology Thailand.
Bjornsson, L., Murto, M., Mattiasson, B. *Microbial Biotechnol.*, 2000, 54, 844–849.
Boe, K., 2006. Online monitoring and control of the biogas process. Ph.D. Thesis, Institute of Environment & Resources, Technical University of Denmark.
Bouallagui, H., Ben Cheikh, R., Marouani, L., Hamdia, M. *Bioresource Technol.*, 2003, 86, 85-89.
Bouallagui, H., Touhami, Y., Ben Cheik, R., Hamdia, M. *Process Biochem.*, 2005, 40, 989-995.
Brown, N.L., 2001. Methane Generation from Human, Animal, and Agricutural Wastes, Books for Business. New York.
Cantrell, K. B., Ducey, T., Ro, K. S., Hunt, P. G. *Bioresource Technol.*, 2008, 99, 7941–7953.
Cecchi, F., Mata-Alvarez, J., 1991, Anaerobic digestion of municipal solid waste; an up-to date review. In: Doelle, H.W., Mitchell, D.A., Rolz, E., ed. Solid substrate cultivation. Amsterdam: Elsevier Applied Science.
Chen, Y., Cheng, J.J., Creamer, S.K. *Biresourse Technol.,* 2007, 99, 4044-4064
Chugh, S., Chynoweth, D.P., Clarke, W., Pullammanappallil, P., Rudolph, V. *Bioresource Technol.,* 1999, 69, 103-115.
De Baere, L.A., Devocht, M., Van Assche, P., Verstraete, W. *Water Resource,* 1984, 18, 543–548.
Dennis, A. And Burke, P.E., 2001. Dairy waste anaerobic digestion handbook. 6007 Hill street Olympia: Environmental energy company
Dinsdale, R.M., Premier, G.C., Hawkes, F.R., Hawkes, D.L. *Bioresource Technol.,* 2000, 72, 159–168.
Edelmann, W., Engeli, H., Gradenecker, M. *Water Sci. Technol.,* 2000, 41, 213–221.
Elango, D., Pulikesi, M., Baskaralingam, P., Ramamurthi, V., Sivanesan, S. *J. hazard. Mater.,* 2007, 41, 301–304.
Erickson, L.E., Fayet, E., Kakumanu, B.K., Davis, L.C., 2004. *Anaerobic Digestion*. National Agricultural Biosecurity Center, Kansas State University.
Fantozzi, F., Buratti, C. *Bioresource Technol.,* 2009, 100, 5783–5789.
Fantozzi, F., Buratti, C. *Bioresource Technol.,* 2011, 102, 8885-8892.
Fernandez, A., Sanchez, A., Font, X. *Biochem. Eng. J.*, 2005, 26, 22–28.
Fernandez, B., Porrier, P., Chamy, R. *Water Sci. Technol.,* 2001, 44, 103-108.
Forster-Carneiro, T., Pérez, M., Romero, L.I. *Bioresource Technol.,* 2008, 99, 6994-7002.
Fricke, K., Santen, H., Wallmann, R., Huttner, A., Dichtl, N. *Waste Management.,* 2007, 27, 30–43.
Garcia-Peña, E.I., Parameswaran, P., Kang, D.W., Canul-Chan, M., Krajmalnik-Brown, R. *Bioresource Technol.,* 2011, 102, 9447-9455.
Gerardi, H.H., 2003. The Microbiology of Anaerobic digesters. John Wiley & Sons Inc, Hoboken. New Jersey.

Grady, C.P.L., Daigger, G.T., Lim, H.C., 1999. BiologicalWaste Water Treatment. Marcel Dekker, New York.

Haleh, S., 2011. Enhancement of Anaerobic Digestion of Organic Fraction of Municipal Solid Waste by Microwave Pretreatment. Ottawa, Canada.

Hartmann, H., Ahring, B.K. *Water Resource*, 2005, 39, 1543–1552.

Hernandez-Berriel, M.C., Benavides, L.M., Perez, D.J.G., Delgado, O.B. *Waste Management*, 2008, 28, 14–20.

Hwang, M.H., Jang, N.J., Hyum, S.H., Kim, I.S. *J. Biotechnol.,* 2004, 111,297–309.

IEA, 1994. Biogas from municipal solid waste; overview of systems and markets for anaerobic digestion of MSW. Report of International Energy Agency Task XI: Conversion of MSW to energy; Activity 4: Anaerobic digestion of MSW.

Igoni, A.H., Ayotamuno, M.J., Eze, C.L., Ogaji, S.O.T., Probert S.D. *Applies Energy*, 2007, 85, 430-438.

Ismail, Z. K., Abderrezaq, S. K. *Energy resource,* 2007, 29, 657–668.

Jin, P., Bhattacharya, S.K., Williama, C.J., Zhang h. *Water Resource*, 1998, 32, 977–988.

Kayhanian, M. *Waste Management Resourse, 1995,* 13, *123-136.*

Kayhanian, M. Environmental Technol., 1999, 20, 355–365.

Khalid, A., Arshad, M., Anjum, M., Mahmood, T., Dawson, L. *Waste Management*, 2011, 31, 1737-1744.

Khanal, S.K., 2008. Overview of Anaerobic Biotechnology.

Kim, H., Shin, H., Han, S., Oh, S. J. *Air & Waste Management Association,* 2007, 57, 309–318.

Kim, J.K., Nhat, L., Chun, Y.N., Kim, S.W. *Biotechnology Bioprocess Eng.,* 2008, 13, 499–504.

Koster, I.W., Rinzema, A., De Vegt, A.L., Lettinga, G. *Water Resource*, 1986, 20, 1561–1567

Krupp, M., Schubert, J., Widmann, R. *Waste Management*, 2005, 25, 393–402

Lahdheb, H., Bouallagui, H., Hamdi, M. *Biosource Technol.,* 2009, 100, 1555-1560.

Lee, J., Song, J., Hwang, S. *Bioresource Technol.,* 2009, 100, 1491–1493.

Macias-Corral, M., Somani, Z., Hanson, A., Smith, G., Funk, P., Yu, H., Longworth, J. *Bioresource Technol.,* 2008, 99, 8288-8293.

Madsen, M., Holm-Nielsen, J.B., Esbensen, K.H., 2011. Monitoring of anaerobic digestion processes: A review perspective [online] Available at: http://www.sciencedirect.com/science/article/pii/S1364032111001705 [Accessed on 21 September 2012].

Mechichi, T, Sayadi, S. *Process Biochemistr.,* 2005, 40, 139–45.

Mohee, R. *Conservation and Recycling*, 2002, 36, 33-43.

MONNET F., 2003. An introduction to Anaerobic Digestion of Organic Wastes [online] Available from: http://www.biogas,ax.co.uk/media/introanaerobicdgestion_073323000_1011_24042007.pdf [Accessed on 18 September 2012].

Mshandete, A., Bjornsson, L., Kivaisi, A.K., Rubindamayugi, M.S.T., Mattiasson, B. *Renewable Energy*, 2006, 31, 2385–2392.

Mudhoo, A., 2012. Biogas Production pretreatment method in Anaerobic Digestion, Scrivener Publishers LLC, USA.

Muller, J., Lehne, G., Schwedes, J., Battenberg, S., Naveke, R., Kopp, J., Dichtl N., Scheminski, A., Krull, R., Hempel, D. C. *Water Science and Technology,* 1998, 38, 425-433.

Muller, W.R., Frommert, I., Jorg, R. *Environmental Sci. Technol.,* 2004, 3, 141-158.
Neves, L., Oliveira, R., Alves, M.M. *Process Biochemistry*, 2004, 39, 2019-24.
Nguyen, P.H.L., Kuruparan, P., Visvanathan, C. *Bioresource Technology,* 2006, 98, 380-387.
O'Flaherty, V., Mahony, T., O'Kennedy, R., Colleran, E. *Process Biochemistry*, 1998a, 33, 555–569.
Poggi-Varaldo, H.M., Valdes, L., Esparza-Garcia, F., Fernandez, V.G. *Water Science and Technol.,* 1997, 35, 197-204.
Raposo, F., Borja, R., Martin, M.A., De La Rubia, A., Rincon, B. *Chemistry Engineering J.,* 2009, 149, 70-77.
RAPPORT, J., ZHANG, R., JENKINS, B.M., WILLIAMS, R.B., 2008. Current Anaerobic Digestion Technologies Used for Treatment of Municipal Organic Solid Waste. [online] Available from: http://www.calrecycle.ca.gov/publications /Organics/2008011.pdf [Accessed on 18 September 2012].
Rehm, H.J, Reed, G, Puhler, A, Stadler, P.J.W., 2000. Biotechnology, vol. 11A: Environmental processes I, 2nd ed. New York: Wiley.
Rocha, E.R., Smith, C.J. *Microbiology*. 2004, 150, 2125-2134.
Sadi, M., 2010, Design and Building og a biogas digester for organic materials gained from solid waste. Thesis (PhD) An-Najah National University Faculty of Graduate Studies, [online] Available from: http://www.najah.edu/thesis/648.pdf [Accessed on 21 September 2012].
Sakar, S., Yetilmezsoy, K., Kocak, E. *Waste Management and Research,* 2009, 27, 13-18.
Siegert, I., Banks, C. *Process Biochemistry*, 2005, 40, 3412–8.
Sunaro, Johari S., Widiasa N.I., Budiyono. *Internat J. of Sci and Eng.,* 2010, Vol 1(2): 41-45.
Turovskiy, Is., Mathai, PK., 2006. Wastewater sludge processing. New York: Wiley.
Vavilin, V.A., Rytov, S.V., Pavlostathis, S,G., Jokela, J., Rintala, J. *Water Sci. and Technol.,* 2003, 48, 147-154.
Ward J.A., Hobbs P.J., Holliman P.J., Jones D.L. *Bioresource Technol.,* 2008, 99, 7928-7940.

In: Anaerobic Digestion
Editor: Ana Torales
ISBN: 978-1-62808-884-7

Chapter 5

ANAEROBIC DIGESTION SUSTAINABILITY EFFECTS ON PLANT-WIDE PERFORMANCE AND AUTOMATION IN CENTRALIZED WASTEWATER TREATMENT

***U. Zaher*[1,*] *and M. Duqqah*[2,†]**

[1]Department of Biological Systems Engineering,
Washington State University, Pullman, WA, US
[2]Department of Land, Water and Environment,
University of Jordan, Amman, Jordan

ABSTRACT

In the absence of globally unified water quality regulations and with the growing sustainability concerns in the water sector, plant-wide modeling is a powerful tool to assist the selection of wastewater treatment technologies, optimize their operations and benchmark their sustainability. This book chapter illustrates these modeling benefits through advanced plant-wide modeling case study to evaluate the plant upgrade with anaerobic digestion (AD) and its effect on plant sustainability and control. We further discuss consequences of supernatant return after AD to the upstream Biological Nitrogen Removal (BNR) and possible technological solutions to avoid the nitrogen overload improving the treatment system sustainability. The International Water Association Benchmark Simulation Model (BSM) was upgraded with Green House Gas models for AD and BNR. Accordingly, the dynamic interactions and the impacts of the AD upgrade on the BNR system could be evaluated. The main interactions were the increased easily degradable substrate and ammonium load to the BNR due to anaerobic degradation of the sludge. This combined effect improved denitrification but increased nitrous oxide emissions and nitrate in the effluent. Aeration control reduced nitrate and nitrite concentrations and nitrous oxide emissions. However nitrous oxide emissions were still

* U. Zaher: Department of Biological Systems Engineering, Washington State University, P.O. Box 646120, Pullman, WA 99164-6120, US. E-mail: mm.duqqah@wsu.edu , zaheru@wsu.edu.

† M. Duqqah: Department of Land, Water and Environment, The University of Jordan, Amman 11942, Jordan. E-mail: mmduqqah@ju.edu.jo.

higher compared to the plant without AD. The water quality impacts remained within the regulated limits for the three treatment scenarios.

AD and aeration control upgrades increased the sustainability of the plant by reducing the sludge production, operational energy, and eutrophication impact by 50%, 70%, and 40%, respectively. The plant carbon footprint of WWT was increased with the AD upgrade due to increased nitrogen load from recycled supernatant and increased N_2O emissions. Reduction of the footprint depends on the post application of the AD effluents. In case of sludge disposal to landfill, AD would reduce the carbon footprint within the boundary of treatment and disposal by 15-20% due to reductions of sludge volumes.

Keywords: Anaerobic digestion, activated sludge, automation, nitrous oxide, green house gases, Plant-wide modeling, sustainability

1. Introduction

Sustainability of wastewater treatment (WWT) takes the three dimensions of water quality, energy efficiency and reduction of Green House Gas (GHG) emissions. Considering these sustainability dimensions contributes to the mitigation of global challenges such as water scarcity, increasing food and energy demands, and climate change. The presented research illustrates a model-based approach for the assessment of the three dimensions in an integrated manner with two main fundamental and applied focal points. The fundamental focus of this chapter is on the Anaerobic Digestion (AD) effect on Biological Nitrogen Removal (BNR) to evaluate impacts on water quality by the detailed modeling of the biological treatment pathways. The applied focus is on the sustainability assessment evaluating sludge production, Green House Gases (GHG), energy efficiency, operation costs, eutrophication and climate change impacts.

AD application to WWT sludge digestion is a proven upgrade to WWTP improving its sustainability within the treatment and disposal boundaries. AD is a closed system from which GHG are controlled and biogas is often utilized for energy generation. However, the AD treatment mechanism is contrasting with aerobic wastewater treatment and, therefore, their integration within the WWTP should be engineered to improve the sought plant-wide sustainability. AD operates under larger organic carbon loading rate of particulate sludges compared to Activated Sludge (AS) that is mainly treating soluble organic carbon load by concentrating it into sludge biomass. AS systems are also designed for BNR by subsequent nitrification and denitrification steps aiming to the ultimate release of ammonium and nitrate loadings as nitrogen gas.

Uncontrolled increase of nitrogen loadings to BNR would increase effluent ammonia and nitrate (increasing eutrophication impacts) but also intermediates such as nitrite (increasing toxicity impacts) and nitrous oxide (increasing GHG and climate change impacts). AD not only releases organic carbon as biogas but also releases concentrated nutrients and easily degradable carbon in its overflow. Conventional solution is recycling these overflows to AS that need to be controlled to avoid the highlighted BNR overload impacts and maintain the WWT plant-wide sustainability.

Advanced modeling is needed for the assessment of nutrients' regulations due to variability in wastewater flows and permitted water quality consents. There is no global structure for water quality regulations since quality consents are so far regional and site specific.

The International Water Association (IWA) developed a set of Activated Sludge (AS) models ASM1, ASM2, ASM2d and ASM3 (Henze et al., 2000) to simulate biological removal of organic carbon, nitrogen and phosphorus. The IWA task group on anaerobic digestion developed the Anaerobic Digestion Model no.1 ADM1 (Batstone et al., 2002) to simulate the stabilization of WWT sludge before disposal or recycling to agriculture. Transformer models were developed to connect different process models in plant-wide simulations of aerobic and anaerobic digestion (AD) unit processes (Zaher et al., 2007) and coupling advanced nitrogen removal processes, e.g. SHARON and Anammox (Volcke et al., 2002). These modeling efforts were collectively tested and integrated with the plant-wide benchmark models BSM1 and BSM2 developed by the IWA task group on benchmarking WWT control strategies (Alex et al., 2011). In this chapter the BSM2 was extended with detailed nitrous oxide emission model to (1) evaluate AD supernatant recycle and subsequent overload effects on nitrification and denitrification, (2) estimate GHG emissions from the biological WWT, and (3) evaluate the overall WWT performance and sustainability.

2. Methods

2.1. BSM2 Simulated Scenarios

Three treatment scenarios were simulated by upgrading the BSM2 model and altering the benchmark plant depicted in Figure 1. The treatment scenarios were altering the plant configuration considering the AS without AD, AS with AD and AS with both AD and aeration control. Influent wastewater characteristics were applied for a typical WWTP in Western Europe based on diurnal and seasonal variations in flows, pollution loads, weather and temperature as defined by Gernaey et al. (2005; 2006) for 609-day dynamic influent after 200-day steady state initialization.

Comparisons of the three treatment scenarios were based on the end year of the dynamic simulation period from day 245 to day 609; July 1st to the subsequent year July 1st.

A set of proven models were used to model physical and biological unit processes in Figure 1 using the indicated Hydraulic Retention Time (HRT) and Solids Retention Time (SRT) design parameters. The primary clarifier was modeled according to Otterpohl and Freund (1992) and Otterpohl et al. (1994). The secondary clarifier was modeled using 8 layer settling model according to Takacs et al. (1991). The thickening and sludge dewatering units were modeled as perfect separators with 98% solids separation efficiency. The storage tank and flow combiners and splitters were modeled on the basis of mass balance with no reactions. ASM1 model was upgraded with the N_2O emission pathways as described in the next section 2.2. The BNR in the AS units was modeled using the ASM1-N_2O upgraded model. The anaerobic digester was modeled using ADM1 (Batstone et al., 2002) and was interfaced with the AS modeling system using transformer models based on (Nopens et al., 2009; Zaher et al., 2007; 2009). The upgraded treatment scenario with aeration control used BSM2 closed loop control applying dissolved oxygen sensor model and proportional-integral (PI) controller model with a dissolved oxygen set-point of 2.0 g O_2/m^3 using the sensor bias level and the default controller model parameters that were defined for BSM2 (Alex et al., 2011).

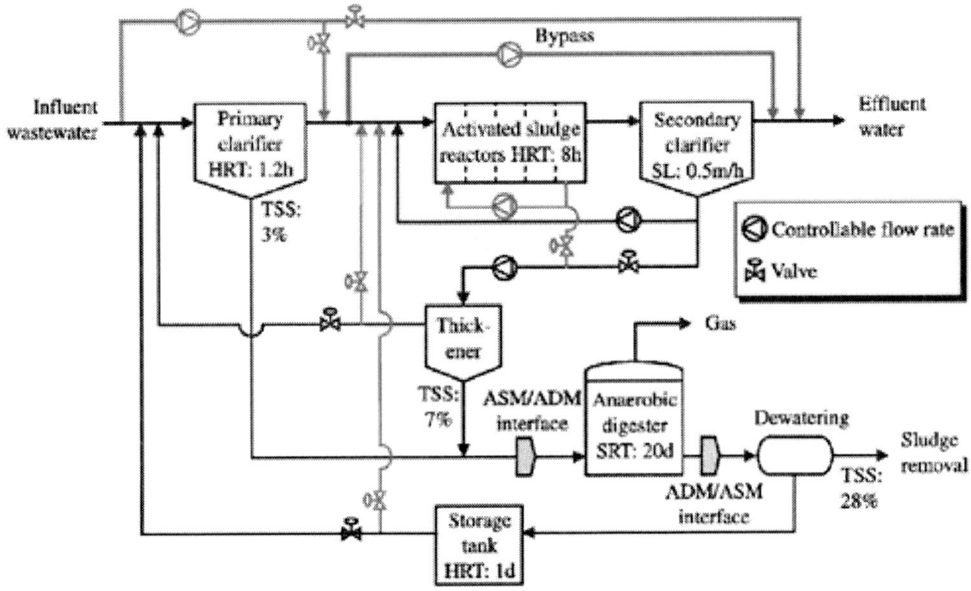

Figure 1. BSM2 benchmark plant (Jeppsson et al., 2007).

2.2. Detailed Modeling of N Removal

Detailed denitrification-nitrification pathways were considered by upgrading ASM1 with N_2O emission pathways under both anoxic and aerobic conditions as listed in Appendix. The upgraded model was implemented in the BSM2 as Matlab-Simulink S function. Denitrification pathway was extended by modeling three steps of anoxic growth of heterotrophic biomass reducing nitrate to nitrite, reducing nitrite to nitrous oxide and reducing nitrous oxide to nitrogen gas.

Easily degradable substrate limitation and inhibitions by oxygen and nitrite were considered in modeling all the anoxic growth steps. Nitric oxide NO is an intermediate of the denitrification pathway but was not considered in the present model upgrade focusing on its successor N_2O as an important GHG. It is hard to track NO compared to N_2O in full-scale applications (Desloover et al., 2012) because of its relatively low production rate and solubility. NO is a toxic gas with a typical 0.2-1 ppm concentration in off-gases from nitrifying-denitrifying AS systems (Kampschreur et al., 2007).

These detected NO concentrations are much below the typical NO exposure limits of 25-30 ppm according to US Occupational Safety and Health Administration (OSHA). If NO emissions exceed these limits NO should be modeled in the denitrification pathway as illustrated elsewhere (Ni et al., 2011; Hiatt and Grady, 2008).

Nitrification was expanded considering two-step growth of autotrophic biomass (Schulthess et al., 1994; Schulthess and Gujer, 1996) and autotrophic nitrous oxide production pathway (Kim et al., 2010; Miyahara et al., 2010). Two autotrophic biomass groups were considered. First is *Nitrosomonas* growth oxidizing ammonia to nitrite and producing two electrons.

The additional capability of *Nitrosomonas* in utilizing the produced electrons in an autotrophic reduction of nitrite to nitrous oxide was modeled according to the recently discovered pathway in (Kim et al., 2010). Second autotrophic group is *Nitrobacter*. Its growth was modeled considering oxidation of nitrite to nitrate.

2.3. Performance and Sustainability Indicators

A set of indicators were applied to compare the studied treatment scenarios. The indicators were classified under two categories of water quality indicators and sustainability indicators.

The water quality indicators were structured into 4 consecutive monitoring levels emulating different regulatory practices world-wide. The primary level (*P*) is for typical monitoring of suspended solids and organic carbon pollution measures setting an upper limit on average concentrations and accounting for the violation time within a year period. The secondary level (*A*) expands the *P* limits to include nutrient concentrations and additionally account for the 95%ile concentration within the year. The ecological level (*E*) expands the evaluation to the concentration of specific compounds whose limits would be different according to the location and sensitivity of receiving water, habitats or planned reuse. Finally, a sustainably level (*S*) is concerned with economic sustainability and given by a formula to calculate water quality index based on the above measures and their treatment cost or penalty structure if violated.

The typical index formula was applied according to equation (1) (Vanrolleghem et al., 1996; Floris-Alsina et al., 2011).

$$EQI = \frac{1}{(t_2 - t_1) \cdot 1000} \int_{t_1=245\ days}^{t_2=609\ days} \begin{pmatrix} B_{SS} \cdot TSS_e(t) + B_{COD} \cdot COD_e(t) + B_{TKN} \cdot TKN_e(t) \\ +B_{NO_x} \cdot NO_{x,e}(t) + B_{BOD5} \cdot BOD_e(t) \end{pmatrix} \cdot Q_e(t) \cdot dt \quad (1)$$

where TSS_e, COD_e, TKN_e, $SNO_{x,e}$ and BOD_e represent, respectively, the total suspended solids, the chemical oxygen demand, the total Kjeldahl nitrogen, the oxidized nitrogen concentration and the biochemical oxygen demand in the effluent. Q_e is the effluent flow rate and t is time. The weights B_x were: $B_{SS} = 2$, $B_{COD} = 1$, $B_{TKN} = 30$, $B_{NOx} = 10$ and $B_{BOD5} = 2$, for the different pollution measures as indicated in subscript.

Sustainability indicators accounted for operational cost, emissions and global impacts. Operational costs were evaluated by the operational cost index (OCI) defined for BSM2 (Jeppsson et al., 2007) as a weighted sum of energy consumption in aeration, mixing, heating, and energy from methane production.

Sustainability analysis considered also sludge production. Emissions and global impacts were evaluated from the average annual results considering both emissions due to effluent discharge and sludge disposal. Global impacts of eutrophication and climate change were evaluated according to Guinée et al. (2002) and ISO, 2006[a,b].

3. Results and Discussions

3.1. N_2O Emissions

Biological production and transport of N_2O were both considered in the N_2O-emmission model implementation and BSM2 upgrade. The dynamics of N_2O emissions are presented in Figure 2 for the simulated three treatment scenarios of AS system only, AS with recycle after AD and the second scenario with aeration control. The fluxes from the aerobic zone (bottom graphs) were 10 folds the fluxes from the anoxic zone due to the higher air flow. Gas stripping rate was modeled proportional to the oxygen transfer rate K_La and temperature. The graphs to lift present a summer to summer year of simulation. More N_2O fluxes from anoxic zones occurred during summer due to higher temperature increasing both heterotrophic denitrification rates and gas stripping. On the contrary, N_2O fluxes from the aerobic zone peaked during winter and lower temperature which has twofold effect. Lower temperature limited the second nitrification step converting NO_2^- to NO_3^- favoring the aerobic pathway for N_2O production from NO_2^-. NO_2^- also peaked during winter. Since applied K_La in the aerobic zone varies between 60-160 d^{-1} compared to 2 d^{-1} in the anoxic zone (just for proper mixing), air flow was enough to stripping the produced N_2O despite lower temperature effects during winter. Similar N_2O diurnal and seasonal dynamics were observed in long-term online monitoring campaign at a full-scale municipal wastewater (Daelman et al., 2013). The study indicated the necessity of increasing measurement frequency and designing sampling times to capture N_2O rapid dynamics.

The prediction of higher N_2O emissions in the aerobic zones is in agreement with recent experimental studies (Miyahara et al., 2010; Kim et al., 2010). These results are also in agreement with the US monitoring survey of N_2O emissions from full scale WWTP's (Ahn et al., 2010). The preliminary results from the survey showed high level of variability due to variations in temperature, pollution load and plant configuration. The survey is therefore expanding with support from the US Water Environment Federation (USWEF) correcting the original US inventory of N_2O emissions that was based on anoxic zones' emissions only (USEPA, 2009). The presented upgrades of the BSM2 model in this chapter account for the different sources of variability and would support such nation-wide surveys.

The model capability of capturing and comparing the variability sources can be illustrated by zooming in the simulation results. The period from 400 day to 450 is zoomed into the right side of Figure 2 to compare dynamics of N_2O emissions from the aforementioned three treatment scenarios. The AS treatment scenario is presented by both the simulated N_2O raw and filtered signals. The raw signal shows the captured dynamics of influent and recycle loads and temperature on hourly basis (four points every hour) in response to metrological and flow dynamics implemented in the model inputs. Other treatments are presented by the filtered signal for clarity. Filtered signals were evaluated using daily average moving window. Same presentation was applied to show the dynamics of other N-components in the next section. Expanding the treatment with AD increased N_2O emissions from both the anoxic and aerobic zones. AD stabilizes sludge releasing nitrogen in ammonium mineralized form which is consequently nitrified and denitrified in the AS plant due to recycle of supernatant from sludge drying and nitrified portion to the anoxic zone upstream the AS plant. In addition, AD produces volatile fatty acids (VFA) to support denitrification with carbon source.

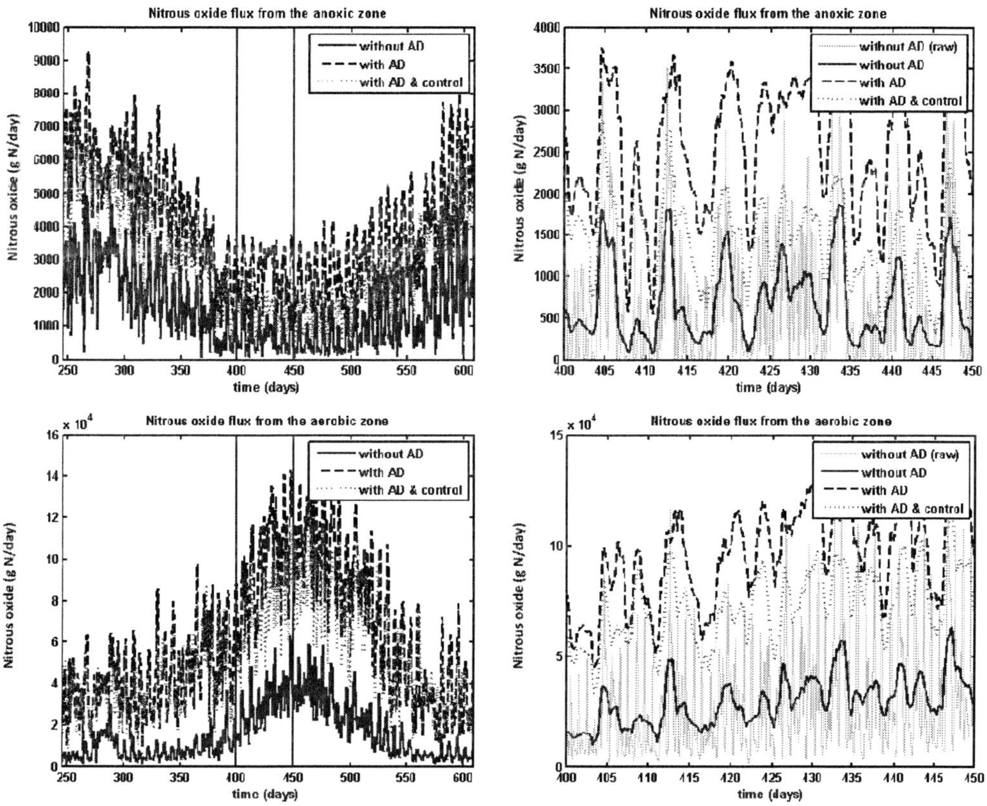

Figure 2. Nitrous oxide emissions from the anoxic (top) and aerobic (bottom) zones for a whole year (left) and the zoomed window (right).

When control is also applied to regulate aeration, nitrate is also reduced and the AD recycled VFA could be used to enhance denitrification.

These improvements can be seen by comparing N_2O emission results from both the anoxic and aerobic zones. These interactions between the AS and AD units and improvements introduced by aeration control were also confirmed by results of other N-components in the next section.

3.2. Dynamics of N State Variables

The dynamics of total effluent ammonium and ammonia (TAN) and effluent nitrate after both the anoxic and aerobic zones are presented in Figure 3. The three graphs on the left show that daily averages were kept under the consented values in Table 1 for a whole year from day 245 to day 609. Zooming into the indicated window between days 400 and 450 into the right side graphs, TAN and nitrate dynamics can be visually analyzed.

Very little differences could be seen between the effluent TAN from the three treatment scenarios indicating efficient aeration and nitrification. Treatment scenario with supernatant recycle after AD and sludge drying caused higher nitrate in the effluent. Applying aeration control lowered nitrate concentrations in the effluent.

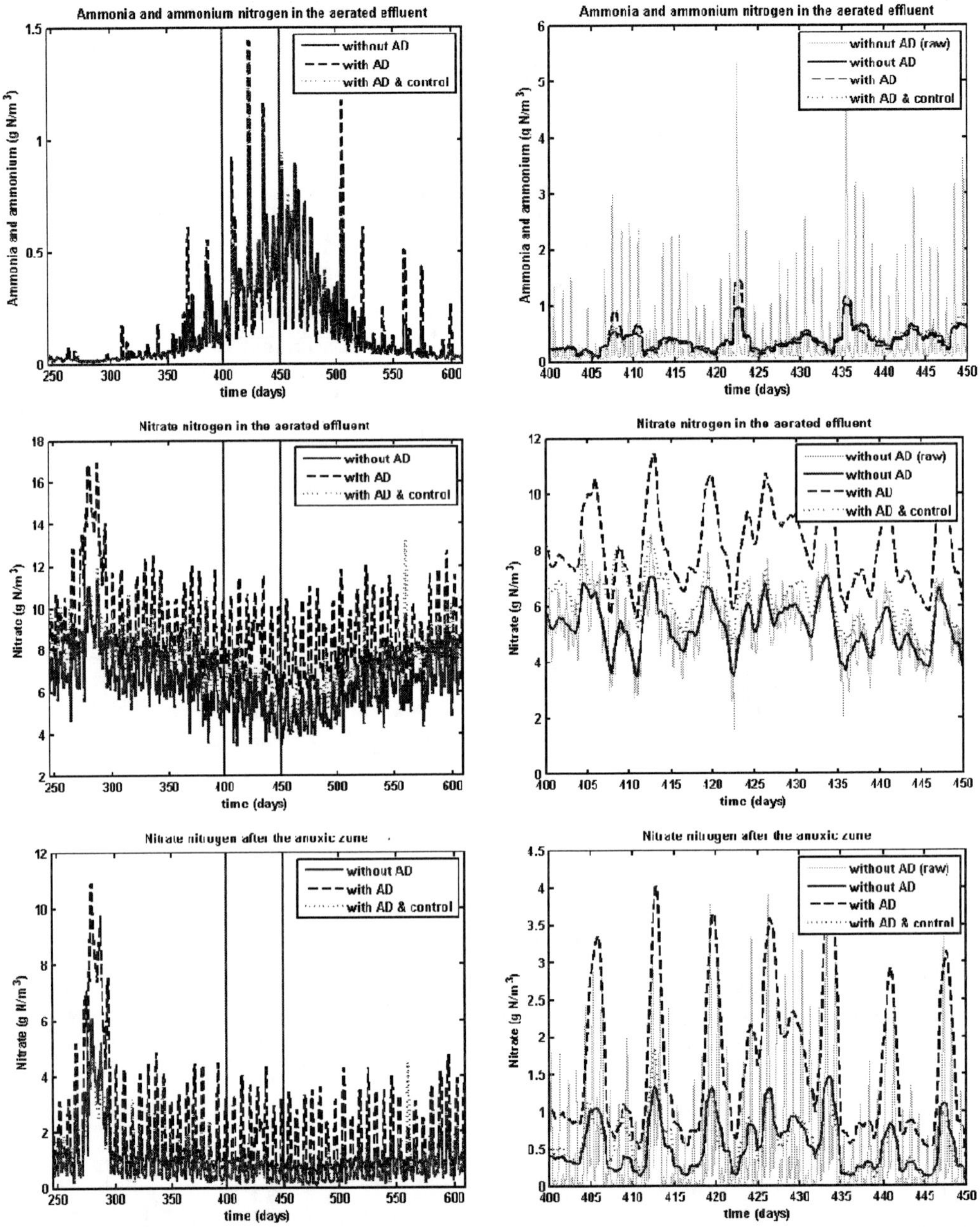

Figure 3. Dynamics of total effluent ammonia and ammonium TAN (top), effluent nitrate (middle) and nitrate after the anoxic zone (bottom) for a whole year (left) and the zoomed window (right).

This improvement is mainly due to improved denitrification. Lowered nitrate concentrations in the effluent and recycled flows to the anoxic zone allowed efficient use of recycled supernatant VFA's and complete denitrification to nitrogen gas. This is confirmed with the simulated nitrate concenterations after the anoxic zone in both AS treatment scenario compared to the treatment scenario with AD and control. The model captured these dynamics by the detailed denitrification modeling considering nitrite production and inhibition to the denitrification chain reactions. Nitrite inhibition was reported in AS experimental studies (Alinsafi et al., 2008).

The upgraded model capability to capture these dynamics would optimize applied solutions to improve effluent quality and nitrogen limits. Practical solutions to improve denitrification are by dosing carbon sources (Lu and Chandran, 2010) that can be substituted by co-digestion (Zaher et al., 2009), e.g. to treat additional feedstock of food waste and benefit from the released VFA. Co-digestion scenarios of sludge (biosolids) and other feedstocks was proven economic (USEPA, 2010). Other solution is to lower TAN loads from sludge drying recycled supernatant by including advanced nitrogen removal processes such as SHARON and Anammox (Volcke et al., 2002). Other solutions is to use more advanced control of the AS system (Flores-Alsina et al., 2011). Decision on such technology selection and upgrade decisions depends on the enforced water quality consents and should be supported with confidence / uncertainty analysis. The presented model would play a significant role in such analysis since it considers and simulates different sources of variability.

In Figure 4, cumulative distribution of total effluent nitrogen (TN) is built from one year simulation results considering seasonal and diurnal variability of pollution loads, flows, and temperature and unit process dynamics and interactions. Accordingly, the maximum effluent TN concentration can be determined at a desired confidence level. As indicated by the dashed lines, the 95% effluent TN concentrations for the three treatment scenarios were 10.9, 15.0 and 13.1 g N/m^3 for AS, AS with AD, and AS, AD and aeration control respectively. These results indicate that adding the sludge AD and treatment would raise the effluent TN but could be maintained under the typical effluent limit 18 g N/m^3 and can be improved by aeration control. A final decision on such upgrade depends on further assessment of WWT sustainability dimensions as discussed in the next section.

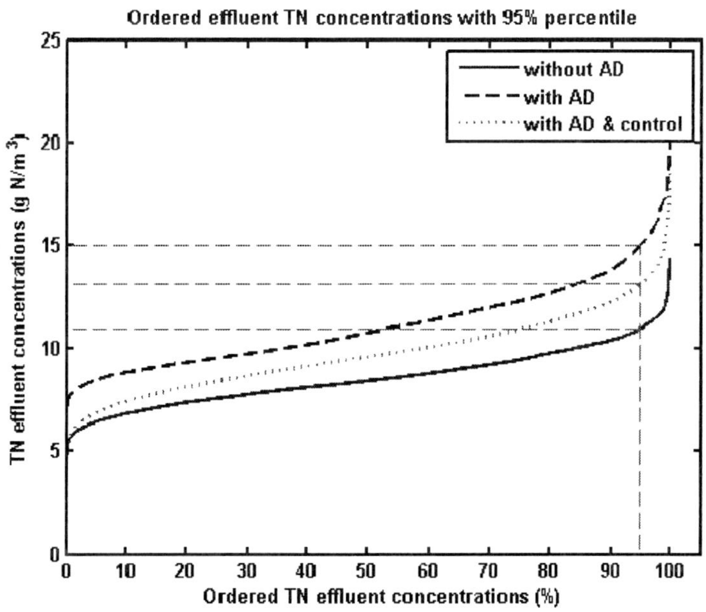

Figure 4. cumulative distribution of the effluent total nitrogen TN.

3.3. WWT Sustainability

The sustainability of the three-studied treatment scenarios was benchmarked using the one-year simulation results between days 245 and 609. The results of the sustainability indicators are discussed in comparison with world-wide regulatory levels and evaluation criteria. The discussions are organized in the following two sections for water quality impacts and other sustainability dimensions.

3.3.1. Water Quality Impacts

As already illustrated in the introduction, there is no world-wide consensus on water quality evaluation criteria. For the sake of clarity, water quality assessment are organized in Table 1 considering the 4 designated regulatory levels P, A, E, S: primary, advanced, ecological and sustainable evaluation. The primary level focuses on effluent clarity and avoidance of septicity of receiving waters. Typical primary monitoring criteria is keeping the average COD, BOD and SS under 100 (g COD/m^3), 10 (g/m^3) and 30 (g SS/m^3), respectively. The primary level is often enforced by penalizing the WWTP's during violation time V_i. It can be seen that there is no significant difference between the studied treatment scenarios on this level.

Table 1. Water quality assessment parameters

Effluent Water quality	Reg.* / limit	Activated Sludge (AS)			AS and Anaerobic digestion (AD)			AS, AD and AS-aeration control		
		A_v**	V_i** (day/ year)	C_{95}**	A_v**	V_i** (day/ year)	C_{95}**	A_v**	V_i** (day/ year)	C_{95}**
Chemical Oxygen Demand COD (g COD/m^3)	P/100	47.9	0.2		48.1	0.2		47.3	0.2	
Biological Oxygen Demand (g/m^3)	P/10	2.42	0.8		2.35	0.8		2.41	0.8	
Suspended solids (g SS/m^3)	P/30	14.4	0.9	18.6	14.5	1.0	18.7	14.9	1.0	18.3
Total ammonium and ammonia (TAN) (g N/ m^3)	A/4	0.2	0.2	0.8	0.2	0.5	0.9	0.2	0.3	0.7
Total Nitrogen (g N/m^3)	A/18	8.5	0.2	10.9	11.0	1.5	15.0	9.8	0.2	13.1
Kjeldahl nitrogen (g N/m^3)	E	2.3			2.2			2.2		
Nitrite nitrogen (g N/m^3)	E	0.1			0.2			0.1		
Nitrate nitrogen (g N/m^3)	E	6.1			8.6			7.4		
Effluent Quality Index (EQI) (kg poll. units/d)	S	4369.4			4901.0			4593.1		

*Regulation levels: Primary (P), Advanced (A) with P included, Ecologically (E) with A included, Sustainably (S) with E included.

**A_v Averaged parameters are based on the average annual load, V_i time of violations, C_{95}: 95 percentile concentration.

The primary water quality standards are applied to WWTP where nutrients loads to receiving waters are mainly coming from other sources such as agriculture. The cumulative effect is however problematic when it comes to the ultimate receiving waters. Thus upgrading to the advanced regulatory level becomes mandatory.

The advanced regulatory level is mainly concerned with eutrophication impacts regulating nutrients such as phosphorous and total nitrogen concentrations and their ratios to prevent outbreak of algal species or water plants. Typical consents on nitrogen components are limiting TN < 18 (g N/m^3) and TAN < 4-5 (g N/m^3). The compliance to these levels is not only monitored according to average concentrations and violation time but also on 95%ile basis. The average TAN results show the same impact from the evaluated treatment scenarios but differences show up considering the 95%ile results. These differences are confirmed when comparing TN results. Again nitrogen limits could be maintained within the regulation limits for the three scenarios with an increase considering AD and possible improvement considering aeration control. Such advanced regulations are typically applied by the European Union (EU) countries to prevent North Sea eutrophication. For instance, a typical TAN consent by Yorkshire Water, UK is 5 (g N/m^3) with statistical monitoring on the 95%le monitoring basis.

In addition, other ecological regulatory limits are set on specific components to reduce impacts to water habitats and fresh water fisheries as well as human health (Vanrolleghem, 2011). For instance, nitrate concentration is regulated under the World Health Organization (WHO) recommendation to be < 50 (g N/m^3) if receiving water is abstracted for potable use to prevent blue babies disease. Also, nitrite and ammonia levels are often regulated to protect fisheries and are regulated according to risk assessment of endangered species. Furthermore, on the sustainablity level, a water quality index is developed to consider the different discharge loads with weighting factors developed on the basis of tariff structure of received pollution loads to the WWTP and/or the cost of treatment. Comparing the ecological limits and the water quality index of the simulated treatment scenarios shows that integrating AD and sludge stabilization increases ecological impacts and water quality index to a small extent. Therefore, other sustainability dimensions are evaluated in the next section.

3.3.2. Sustainability Dimensions

The sustainability of upgrading the WWT with AD and aeration control is evaluated by comparing sludge production, green house gases (GHG), energy consumption and production, operational cost and global impacts to eutrophication and climate change of the three treatment scenarios in Table 2.

The quantity of sludge was reduced by > 50% with the suggested upgrades. The sludge was also stabilized by reducing its COD and N content by 56% and 53%, respectively. The reduced quantity reduces the disposal cost while stabilization increases the opportunity of utilizing the sludge as an organic fertilizer. Comparing GHG emissions, CO_2 from AS was at the same level for the three treatment scenarios. Upgrading with AD increased N_2O emissions 3 times due to the recycled TAN to the AS plant. Considering sludge disposal to landfill and the associated methane emissions were evaluated from the COD balance and by considering complete mineralization of the sludge. Methane from digested sludge was about half the methane production from land filling of undigested sludge since methane was captured first during the digestion process. Comparing the energy uses, aeration control saves 20% of the aeration energy.

Table 2. Evaluation of sustainability dimensions upgrading biological WWT with anaerobic digestion and aeration control

Sustainability dimension	Activated Sludge (AS)	ASand Anaerobic digestion (AD)	AS, AD and AS-aeration control
Sludge			
Production (t Dry solids/year)	2,076	973	978
Total COD (t COD/year)	2,632	1,157	1,170
Total nitrogen (kg N/year)	148,920	70,080	71,175
Green House Gases (GHG)			
CO_2 from aerobic treatment (t/year)	5,647	5,639	5,645
CH_4 from sludge landfill (t/year)	658	289	292
N_2O from aerobic treatment (t/year)	9.4	36.4	30.9
Energy			
Aeration energy consumption (MWh/Year)	1,859	1,859	1,489
Pumping energy consumption (MWh/Year)	161	161	162
Mixing energy consumption (MWh/Year)	280	280	280
Heating energy consumption (MWh/Year)	-	1,469	1,480
Methane energy production (MWh/Year)	-	5,079	5,158
Operational cost			
Operational Cost Index	22,520	7,412	6,394
Global impacts: from disposal and discharge			
Eutrophication (kg PO_4^{3-} eq /year)	149,076	92,213	88,499
Climate Change (t CO_2 eq /year)	22,259.70	18,943.18	17,316.45

Energy production from the AD methane was 3.5 time the heating requirement of the digesters. Evaluating the operational cost index that is mainly a function of different energy uses and sludge production, the AD save about 70%.

The global impacts were evaluated using broader boundaries across the WWTP fence. Eutrophication was evaluated considering both effluent discharges and sludge disposal. Eutrophication impacts were reduced by 40% due to the upgrade with AD and were further improved by considering aeration control.

Climate change was evaluated considering emissions from biological treatment, disposal and energy consumption. Reduction of the carbon footprint within the boundaries of WWTP-landfill was reduced by 15% due to the upgrade with AD and 20% with additional upgrade of aeration control.

CONCLUSION

The presented detailed wastewater treatment plant-wide (WWTP) modeling considered the variability of flows, pollution loads, and temperature and unit process interactions. Consideration of these dynamics allowed benchmarking of effluent water quality according to different levels of regulatory consents.

Furthermore, detailed modeling of unit processes in an integrated manner allowed the sustainability assessment of the whole wastewater treatment plant and its upgrades.

The suggested upgrade with sludge anaerobic digestion (AD) increased nitrous oxide emissions and effluent total nitrogen concentrations. However, with efficient and controlled aeration effluent water quality was maintained within the consented limits considering the different regulatory levels. The increased nitrous oxide emissions with the upgrade were offset by the energy production from the AD methane gas and reduced emissions from sludge disposal to landfill. The suggested upgrade reduced energy use, eutrophication and climate change impacts increasing the sustainability of the whole WWTP and post reuse/recycling.

APPENDIX

N2O Emission Model Extension to the IWA –ASM1 Model (Reaction Rates)

Reaction #	ASM1	N2O model	Process Rate
1	Aerobic growth of heterotrophic biomass		$\mu_{\max H} \cdot \frac{S_S}{K_S + S_S} \cdot \frac{S_O}{K_{OH} + S_O} \cdot X_{BH}$
2.1	Anoxic growth of heterotrophic biomass	Nitrate reduction	$\eta_g \cdot \mu_{\max H} \cdot \frac{S_S}{K_S + S_S} \cdot \frac{K_{OH}}{K_{OH} + S_O} \cdot \frac{K_{NO2,1}}{K_{NO2,1} + S_{NO2}} \cdot \frac{S_{NO3}}{K_{NO3H} + S_{NO3}} \cdot X_{BH}$
2.2		Nitrite reduction	$\eta_g \cdot \mu_{\max H} \cdot \frac{S_S}{K_S + S_S} \cdot \frac{K_{OH}}{K_{OH} + S_O} \cdot \frac{K_{NO2,2}}{K_{NO2,2} + S_{NO2}} \cdot \frac{S_{NO2}}{K_{NO2H} + S_{NO2}} \cdot X_{BH}$
2.3		N_2O Reduction	$\eta_g \cdot \mu_{\max H} \cdot \frac{S_S}{K_S + S_S} \cdot \frac{K_{OH}}{K_{OH} + S_O} \cdot \frac{K_{NO2,3}}{K_{NO2,3} + S_{NO2}} \cdot \frac{S_{N2O}}{K_{N2OH} + S_{N2O}} \cdot X_{BH}$
3.1	Growth of Autotrophs biomass	Nitrite reduction (low oxygen growth of Nitrosomonas)	$\mu_{\max NS} \cdot \frac{S_{NH}}{K_{NH} + S_{NH}} \cdot \frac{S_{NO2}}{K_{NO2NS} + S_{NO2}} \cdot X_{BNS}$
3.2		Nitrosomonas	$\mu_{\max NS} \cdot \frac{S_{NH}}{K_{NH} + S_{NH}} \cdot \frac{S_O}{K_{ONS} + S_O} X_{BNS}$
3.3		Nitrobacter	$\mu_{\max NB} \cdot \frac{S_O}{K_{ONB} + S_O} \cdot \frac{S_{NO2}}{K_{NO2NB} + S_{NO2}} \cdot X_{BNB}$
4	Decay of heterotrophic biomass		$b_H \cdot X_{BH}$
5.1	Decayof Nitrosomonas		$b_{NS} \cdot X_{BNS}$
5.2	Decayof Nitrobacter		$b_{NB} \cdot X_{BNB}$
6	Ammonification of soluble organic nitrogen		$k_a \cdot S_{ND} \cdot X_{BH}$
7	Hydrolysis of slowly biodegradable substrate		$k_h \cdot \frac{X_S / X_{BH}}{K_X + X_S / X_{BH}} \cdot \left[\frac{S_O}{K_{OH} + S_O} + \eta_h \frac{K_{OH}}{K_{OH} + S_O} \cdot \frac{S_{NO2}}{K_{NO2H} + S_{NO2}} \cdot \frac{S_{NO3}}{K_{NO3H} + S_{NO3}} \right] \cdot X_{BH}$
8	Hydrolysis of Organic Ntrogen		$k_h \cdot \frac{X_{ND} / X_{BH}}{K_X + X_{ND} / X_{BH}} \cdot \left[\frac{S_O}{K_{OH} + S_O} + \eta_h \frac{K_{OH}}{K_{OH} + S_O} \cdot \frac{S_{NO2}}{K_{NO2H} + S_{NO2}} \cdot \frac{S_{NO3}}{K_{NO3H} + S_{NO3}} \right] \cdot X_{BH}$

N_2O Emission Model Extension to the IWA –ASM1 Model (Stoichiometry)

Reaction #	ASM1 Component #		1	2	3	4	5	6		7	8	9				10	11	12	13
	N_2O Component#							6.1	6.2			9.1	9.2	9.3					
	ASM1		S_I	S_S	X_I	X_S	X_{BH}	X_{BA}		X_P	S_O	S_{NO}				S_{NH}	S_{ND}	X_{ND}	S_{ALK}
		N2O model						X_{BNS}	X_{BNB}			S_{NO3}	S_{NO2}	S_{N2O}	N_2				
		Units	g COD/m^3									g N/m^3							equ/m^3
1	Aerobic growth of heterotrophic biomass			$\frac{-1}{Y_H}$			1				$\frac{-1+Y_H}{Y_H}$					$-i_{XB}$			$\frac{-i_{XB}}{14}$
2.1	Anoxic growth of heterotrophic biomass	Nitrate reduction		$\frac{-1.14-Y_H}{Y_H}$			1					$\frac{-1}{Y_H}$	$\frac{1}{Y_H}$			$-i_{XB}$			$\frac{-i_{XB}}{14}$
2.2		Nitrite reduction		$\frac{-1.14-Y_H}{Y_H}$			1						$\frac{-1}{Y_H}$	$\frac{1}{Y_H}$		$-i_{XB}$			$\frac{-2}{14 \cdot Y_H}-\frac{i_{XB}}{14}$
2.3		N_2O Reduction		$\frac{-1}{Y_H}$			1							$\frac{-1+Y_H}{0.57Y_H}$		$-i_{XB}$			$\frac{-i_{XB}}{14}$
3.1	Growth of Autotrophs biomass	Nitrite reduction (low oxygen growth of Nitrosomonas)						1					$\frac{-2.28+Y_{NS}}{3.43 \cdot Y_{NS}}$	$\frac{1}{Y_{NS}}$		$\frac{-1.14+Y_{NS}}{3.43 \cdot Y_{NS}}-i_{XB}$			$\frac{1.14}{14 \cdot 3.43 \cdot Y_{NS}}-\frac{i_{XB}}{14}$
3.2		Nitrosomonas						1			$\frac{-3.43+Y_{NS}}{Y_{NS}}$		$\frac{1}{Y_{NS}}$			$-i_{XB}-\frac{1}{Y_{NS}}$			$-\frac{2}{14 \cdot Y_{NS}}-\frac{i_{XB}}{14}$
3.3		Nitrobacter							1		$\frac{-1.14+Y_{NB}}{Y_{NB}}$	$\frac{1}{Y_{NB}}$	$\frac{-1}{Y_{NB}}$			$-i_{XB}$			$\frac{-i_{XB}}{14}$
4	Decay of heterotrophic biomass					$1-f_P$	-1			f_P								$i_{XB}-f_P \cdot i_{XP}$	
5.1	Decayof Nitrosomonas					$1-f_P$		-1		f_P								$i_{XB}-f_P \cdot i_{XP}$	
5.2	Decayof Nitrobacter					$1-f_P$			-1	f_P								$i_{XB}-f_P \cdot i_{XP}$	
6	Ammonification of soluble organic nitrogen															1	-1		$\frac{1}{14}$
7	Hydrolysis of slowly biodegradable substrate			1		-1													
8	Hydrolysis of Organic Ntrogen																1	-1	

N_2O EMISSION MODEL EXTENSION TO THE IWA –ASM1 MODEL (SYMBOLS AND PARAMETERS)

State Variables		S_S	Soluble organic substrate g COD/ m^3
S_{ALK}	Alkalinity equ/m^3	X_{BA}	Total autotrophic bacteria g COD/ m^3
S_I	Soluble inert g COD/m^3	X_{BH}	Total heterotrophic bacteria g COD/ m^3
S_{N2O}	Nitrous oxide g N/m^3	X_{BNB}	Total Nitrobacter g COD/ m^3
S_{ND}	Soluble organic nitrogen g N/m^3	X_{BNS}	Total Nitrosomonas g COD/ m^3
S_{NH}	Total Ammonium and ammonia g N/ m^3	X_I	Particulate inert g COD/ m^3
S_{NO2}	Nitrite g N/ m^3	X_{ND}	Particulate organic nitrogen g N/ m^3
S_{NO3}	Nitrate g N/ m^3	X_P	Particulate decaying biomass g COD/ m^3
S_O	Oxygen g COD/ m^3	X_S	Slowly biodegradable substrate g COD/ m^3

Stoichiometric Parameters		Unit	value
Y_H	Heterotrophic yield	-	0.67
Y_{NS}	Autotrophic yield of Nitrosomonas	-	0.21
Y_{NB}	Autotrophic yield of Nitrobacter	-	0.04
i_{XB}	Nitrogen content of biomass	g N/g COD	0.086
i_{XP}	Nitrogen content of particulate	g N/g COD	0.06
f_P	Fraction of biomass yielding particulate products		0.08

Kinetic Parameters			
μ_{maxH}	Maximum specific growth rate of Heterotrophs	day^{-1}	4.0
μ_{maxNS}	Maximum specific growth rate of Nitrosomonas	day^{-1}	0.86
μ_{maxNB}	Maximum specific growth rate of Nitrobacter	day^{-1}	1.26
η_g	Correction factor of anoxic growth rate of Heterotrophs	-	0.8
η_h	Correction factor of anoxic hydrolysis	-	0.8
b_H	Heterotrophic decay rate	day^{-1}	0.3
b_{NS}	Decay rate of Nitrosomonas	day^{-1}	0.05
b_{NB}	Decay rate of Nitrobacter	day^{-1}	0.05
k_a	Ammonification rate	m^3/g COD/day	0.05
k_h	Maximum specific hydrolysis rate	day^{-1}	3.0
K_X	Half saturation coefficient for hydrolysis	-	0.1
K_S	Half saturation coefficient for Heterotrophs	g COD/m^3	10.0
K_{OH}	Oxygen half saturation coefficient for Heterotrophs	g O_2/m^3	0.2
K_{NO3H}	Nitrate half saturation coefficient for Heterotrophs	g N/m^3	0.25
K_{NO2H}	Nitrite half saturation coefficient for Heterotrophs	g N/m^3	0.81
K_{N2OH}	Nitrous oxide half saturation coefficient for Heterotrophs	g N/m^3	0.005
$K_{NO2,1}$	Nitrite inhibition constant for nitrate reduction	g N/m^3	12.0
$K_{NO2,2}$	Nitrite inhibition constant for nitrite reduction	g N/m^3	11.31
$K_{NO2,3}$	Nitrite inhibition constant for nitrous oxide reduction	g N/m^3	17.00

Kinetic Parameters			
K_{NH}	Amonium half saturation coefficient for Nitrosomonas	g N/m^3	1.0
K_{NO2NS}	Nitrite half saturation coefficient for Nitrosomonas	g N/m^3	18.81
K_{NO2NB}	Nitrite half saturation coefficient for Nitrobacter	g N/m^3	0.59
K_{ONS}	Oxygen half saturation coefficient for Nitrosomonas	g O_2/m^3	0.5
K_{ONB}	Oxygen half saturation coefficient for Nitrobacter	g O_2/m^3	0.59

ACKNOWLEDGMENTS

We appreciate the efforts and contributions made by the International Water Association – Task Group on Benchmarking Wastewater Treatment Control Strategies in developing the Benchmark Simulation Model no. 2 (BSM2). Special thanks are due to Dr. Ulf Jeppsson, head of Industrial Electrical Engineering and Automation (IEA) department, Lund University, Sweden, for the Matlab-Simulink implementation of BSM2.

REFERENCES

Ahn, J. H., Kim, S., Park, H., Rahm, B., Pagilla, K., and Chandran, K. (2010) N_2O Emissions from activated sludge processes, 2008–2009: results of a national monitoring survey in the United States, *Environmental Science and Technology*, 44, 4505-4511.

Alex, J., Amerlinck, Y., Benedetti, L., Comas, Q., Copp, J., Corominas, L., Gernaey, K. V., Jeppsson, U., Nopens, I., Pons, M. N., Rodriguez-Roda, I., Rosen, C., Steyer, J. P., Vanrolleghem, P. A., Volcke, E., and Vrecko, D. (2011). *Benchmarking of Control Strategies for Wastewater Treatment Plants*. Editors: Gernaey, K. V., Jeppsson, U., Vanrolleghem, P. A., Copp, J. B., and Steyer, J. P. Scientific and Technical Report, IWA Publishing, London, U.K, (ISBN: 9781843391463).

Alinsafi, A., Adouani, N., Beline, F., Lendormi, T., Limousy, L., and Sire, O. (2008) Nitrite effect on nitrous oxide emission from denitrifying activated sludge, *Process Biochemistry*, 43, 683-689.

Batstone, D. J., Keller, J., Angelidaki, I., Kalyuzhnyi, S. V., Pavlostathis, S. G., Rozzi, A., Sanders, W. T. M., Siegrist, H., and Vavilin, V. A. (2002). *Anaerobic Digestion Model No. 1. IWA STR No 13,* IWA Publishing, London, UK.

Daelman, M. R. J., Baets, B. D., van Loosdrecht, M. C. M., and Volcke, E. I. P. (2013) Influence of sampling strategies on the estimated nitrous oxide emission from wastewater treatment plants, *Water Research*, 47, 3120-3130.

Desloover, J., Vlaeminck, S. E., Clauwaert, P., Verstraete, W., and Boon, N. (2012) Strategies to mitigate N_2O emissions from biological nitrogen removal systems, *Current Opinion in Biotechnology*, 23, 474–482.

Flores-Alsina, X., Corominas, L., Snip, L., Vanrolleghem, P. A. (2011) including greenhouse gas emissions during benchmarking of wastewater treatment plant control strategies, *water research* 45, 4700-4710.

Gernaey, K. V., Rosen, C. and Jeppsson, U. (2006) WWTP dynamic disturbance modelling - an essential module for long-term benchmarking development, *Water Science and Technology,* 53(4–5), 225–234.

Gernaey, K. V., Rosen, C., Benedetti, L., and Jeppsson, U. (2005) Phenomenological modelling of wastewater treatment plant influent disturbance scenarios, *Preprints of 10th International Conference on Urban Drainage (10ICUD)*, 21–26 August, Copenhagen, Denmark.

Guinée, J. B., Gorrée, M., Heijungs, R., Huppes, G., Kleijn, R., Koning, A. de, Oers, L. van, Wegener Sleeswijk, A., Suh, S., Udo de Haes, H. A., Bruijn, H. de, Duin, R. van, and Huijbregts, M. A. J. (2002) *Handbook on life cycle assessment.* Operational guide to the ISO standards. I: LCA in perspective. IIa: Guide. IIb: Operational annex. III: Scientific background. Kluwer Academic Publishers, ISBN 1-4020-0228-9, Dordrecht, 2002, 692 pp.

Henze, M., Gujer, W., Mino, T., and van Loosdrecht, M. C. M. (2000). Activated Sludge Models: ASM1, ASM2, ASM2d and ASM3, IWA Publishing, London, UK. *Scientific and Technical Report No: 9.*

Hiatt, W. C. and Grady, C. P. L. Jr. (2008) An Updated Process Model for Carbon Oxidation, Nitrification, and Denitrification, *Water Environment Research*, 80(11), 2145-2156.

ISO (2006[a]) ISO 14040 International Standard. In: Environmental Management –Life Cycle Assessment – Principles and Framework. International Organisation for Standardization, Geneva, Switzerland.

ISO (2006[b]) ISO 14044 International Standard. In: Environmental Management –Life Cycle Assessment – Requirements and Guidelines. International Organisation for Standardisation, Geneva, Switzerland.

Jeppsson, U., Pons, M.-N., Nopens, I., Alex, J., Copp, J. B., Gernaey, K. V., Rosen, C., Steyer, J.-P., and Vanrolleghem, P. A. (2007) *Benchmark simulation model no 2: general protocol and exploratory case studies*, 56(8), 67-78.

Kampschreur, M. J., Picioreanu, C., Tan, N., Kleerebezem, R., Jetten, M. S. M., and van Loosdrecht, M. C. M. (2007) Unraveling the Source of Nitric Oxide Emission During Nitrification, *Water Environment Research*, 79(13), 2499-2509.

Kim, S., Miyahara, M., Fushinobu, S, Wakagi, T., and Shoun, H. (2010) Nitrous oxide emission from nitrifying activated sludge dependent on denitrification by ammonia-oxidizing bacteria, *Bioresource Technology*, 101, 3958-3963.

Lu, H. and Chandran, K. (2010) Factors promoting emissions of nitrous oxide and nitric oxide from denitrifying sequencing batch reactors operated with methanol and ethanol as electron donors, *Biotechnology and Bioengineering*, 106(3), 390-398.

Miyahara, M., Kim, S., Fushinobu, S., Takaki, K., Yamada, T., Watanabe, A., Miyauchi, K., Endo, G., Wakagi, T., and Shoun, H. (2010) Potential of aerobic denitrification by *Pseudomonas stutzeri* TR2 to reduce nitrous oxide emissions from wastewater treatment plants, *Applied and Environmental Microbiology*, 76(14), 4619-4625.

Ni, B-J., Ruscalleda, M., Pellicer-Nacher, C., and Smets, B. F. (2011) SmetsModeling Nitrous Oxide Production during Biological Nitrogen Removal via Nitrification and Denitrification: Extensions to the General ASM Models, *Environmental Science and Technology,* 45, 7768–7776.

Nopens, I., Batstone, D. J., Copp, J. B., Jeppsson, U., Volcke, E., Alex, J., and Vanrolleghem, P. A. (2009) An ASM/ADM model interface for dynamic plant-wide simulation, *Water Research,* 43, 1913-1923.

Otterpohl, R. and Freund, M. (1992) Dynamic models for clarifiers of activated sludge plants with dry and wet weather flows, *Water Science and Technology*, 26(5–6), 1391–1400.

Otterpohl, R., Raak, M. and Rolfs, T. (1994) A mathematical model for the efficiency of the primary clarification, *IAWQ 17th Biennial International Conference*, 24–29 July, Budapest, Hungary.

Schulthess, R. V. and Gujer, W. (1996) Release of nitrous oxide (N2O) from Denitrifying activated sludge: verification and application of a mathematical model, *Water Research*, 30 (3), 521-530.

Schulthess, R. V., Wild, D. and Gujer, W. (1996) Nitric and nitrous oxides from denitrifying activated sludge at low oxygen concentration, *Water Science and Technology*, 30(6), 123-132.

Takacs, I., Patry, G. G. and Nolasco, D. (1991) A dynamic model of the clarification thickening process, *Water Research*, 25(10), 1263–1271.

USEPA (2009) Inventory of US Greenhouse Gas Emissions and Sinks: 1990-2006, EPA report no. 430-R-08-005; US Environmental Protection Agency: Washington, DC.

USEPA (2010) Co-Digestion Economic Analysis Tool (CoEAT), Pacific Southwest, Region 9, http://www.epa.gov/region9/organics/coeat/index.html

Vanrolleghem, P. A. (2011) Resolving the tensions over the treatment plant's fence: Diverse approaches around the world. *Lecture held at the WEF/IWA Nutrient Recovery and Management Conference 2011.* Miami, FL, US, January 9-12, 2011.

Vanrolleghem, P. A., Jeppsson, U., Carstensen, J., Carlsson, B., and Olsson, G. (1996) Integration of WWT plant design and operation - A systematic approach using cost functions, *Water Science and Technology*, 34(3-4), 159-171.

Volcke, E. I. P., Hellinga, C., Van Den Broeck, S., van Loosdrecht, M. C. M., and Vanrolleghem, P. A. (2002) Modelling the SHARON process in view of coupling with Anammox. *Proceedings 1st IFAC TiASWiK Conference*. Gdansk-Sobieszewo, Poland, June 19-21 2002, 65-72.

Zaher, U., Grau, P., Benedetti, L., Ayesa, E., Vanrolleghem, P. A. (2007) Transformers for interfacing anaerobic digestion models to pre- and post-treatment processes in a plant-wide modelling context. *Environmental Modelling and Software* 22(1), 40–58.

Zaher, U., Li, R., Jeppsson, U., Steyer, J.-P., and Chen, S. (2009) GISCOD: general integrated solid waste co-digestion model, *Water Research*, 43(10), 2717-2727.

In: Anaerobic Digestion
Editor: Ana Torales

ISBN: 978-1-62808-884-7

Chapter 6

ANAEROBIC CO-DIGESTION OF CHEESE WHEY AND THE SCREENED LIQUID FRACTION OF DAIRY MANURE: BATCH TESTS WITH DIFFERENT CHEESE WHEY PROPORTIONS

Carlos Rico[1*], José Luis Rico[2] and Noelia Muñoz[2]

[1] Department of Water and Environmental Science and Technologies, University of Cantabria. Avda. Los Castros s/n, Santander, Spain

[2] Department of Chemical Engineering and Inorganic Chemistry, University of Cantabria. Avda. Los Castros s/n, Santander, Spain

ABSTRACT

Anaerobic co-digestion of animal manures and wastes or by-products from the food industry has become a key method for both waste treatment and the production of renewable fuels. In co-digestion processes, the proportion of substrates plays an important role due to synergies amongst them. However, the proportions employed must be carefully studied to avoid inhibitions due to toxic compounds or acidification. In this work, anaerobic co-digestion of cheese whey and screened liquid fraction of dairy manure has been investigated under batch mode operation. Six cheese whey proportions were tested, from 10 to 60%. Dairy manure and cheese whey resulted to be a good match for anaerobic co-digestion. Co-digestion of both substrates permitted to digest a difficult substrate such as the whey without alkalinity addition. However, above 20% cheese whey proportion, methane yield was lower than expected from methane potentials of both substrates separately. The best results in terms of methane conversion were obtained at 20% whey proportion. Under this condition methane yield of the mixture was 46.4% higher than that of the liquid manure, and cheese whey was completely degraded with an estimate methane yield of 18.4 L CH_4 kg^{-1} whey. On the contrary, with 60% cheese whey proportion, the methane yield was 82.7% higher than that of the liquid manure but the estimate methane yield of the whey was reduced to 13.2 L CH_4 kg^{-1}. Alkalinity of dairy

[*] Corresponding author. Tel.: +34-942201848; fax: +34-942201703; E-mail address: ricoc@unican.es

manure was enough to develop the anaerobic co-digestion process. However, above 20% cheese whey proportion, methane yields were lower than expected.

Keywords: Cheese whey; Co-digestion; Dairy manure; Methane; Batch test

1. INTRODUCTION

The increase in production and concentration of intensive livestock operations have resulted in greater awareness and concern for the proper storage, treatment and utilization of livestock manure. Much larger dairy farms have become more common since 1990. To collect and store manure at these large farms, farmers often use liquid manure management systems that use water to flush or clean alleyways or pits where the manure is excreted. This liquid and manure mixture is generally collected and stored until it can be applied to cropland using irrigation equipment. While in storage, the submerged manure generates methane. Moreover, typical disposal methods for animal manure allow for the emission of ammonia, particulate matter, unpleasant odours, volatile organic compounds and a variety of other air pollutants, which can damage the environment and pose risks to animal and human health [1, 2]. In addition when manure is not properly managed can cause severe environmental problems such as eutrophication of water bodies due to its high organic matter, nitrogen and phosphorous concentrations [3, 4].

Cheese whey is the major by-product of dairy industries and presents rather high pollutant characteristics. It is produced in high amounts which are constantly increasing [5]. The composition of cheese whey depends on a lot of factors such as the composition and quality of the milk, production techniques, the amount of yeast or acid used and the period of coagulation and temperature [6]. Typically, cheese whey has 5-8% of dry matter of which around 10–20% are proteins, 60–80% are lactose and the rest are minerals, vitamins, fat, lactic acid and trace elements [7]. Cheese whey presents high organic load (up to 80 g COD L^{-1}), very low alkalinity content and high biodegradability [8]. Cheese whey can be processed in the dairy industry to obtain valuable products as lactoproteins and lactose, or directly used as animal feed [9-11]. Production problems of size and location of dairy factories makes collection and shipment of cheese whey to industrial plants for processing and production of proteins and lactose not profitable. In the past, most of the cheese factories disposed their effluents by land application or direct discharge to receiving waters (rivers, lakes, ocean, etc.) without any pre-treatment [12]. Approximately 47% of the 115 million tons of whey produced world-wide every year are disposed of in the environment [13]. Because of its high organic load, cheese whey disposal constitutes a serious environmental problem. Anaerobic digestion of cheese whey has been reported by various researchers [14-18]. Although this high biodegradable organic load is a chance for both green energy generation and pollution control through anaerobic digestion, cheese whey is a quite problematic substrate for being treated by anaerobic process due to the lack of alkalinity, the high COD concentration and the tendency to acidify very rapidly [19]. Low methane yields have been associated with the low pH of the fermented whey [16]. Furthermore, some authors reported about external nutrients supplementation to maintain system stability [14, 20].

Anaerobic co-digestion of animal manure and by-products from agriculture and the food industry demonstrates many advantages. Biogas plants can convert a disposal problem into a profit center, allowing animal manures, food wastes and agricultural crops to be converted into a highly valuable fuel, reducing greenhouse gases emissions and replacing mineral fertilization by nutrient recovery [21-24]. Therefore, anaerobic digestion has become a key method for both waste treatment and the production of renewable fuels [25, 26]. Co-digestion of different substrates may enhance the anaerobic digestion process due to better carbon and nutrient balance [27] and is a suitable option for improving biogas production [28].

The characteristics of dairy manure and cheese whey are completely opposite. Dairy manure presents many suspended solids and fibrous material and only a small part of the organic matter content is in soluble form. Manure has enough alkalinity and nutrients to develop the anaerobic process and its anaerobic biodegradability is about 45%. Hydrolysis is the rate-limiting step in the anaerobic digestion of manure whereas for the whey it is the methanogenesis. Manure and cheese whey are totally antagonistic but complementary and therefore have to be a good match for a co-digestion process as reported by Gelegenis et al. [29]. To date very few attempts have been carried out for co-digestion of cow manure and cheese whey [6, 30-32].

In this work the anaerobic co-digestion of cheese whey and the screened liquid fraction of dairy manure has been studied in batch tests at mesophilic conditions. The tests were conducted under six different cheese whey proportions, from 10 to 60% in order to evaluate the influence of cheese whey proportion in methane production and stability of the process.

2. Materials and Methods

2.1. Substrates

Cheese whey (CW) was received from a dairy company, Queserías la Fuente, located in Heras (Cantabria, Spain, Lat: N 43°23′28′′ Long: W 3°45′46′′). CW was collected and immediately delivered to the laboratory, where it was stored at 4°C prior to use. Storage time was less than 48 hours before samples were processed further. Characteristics of CW can be found in Table 1.

Table 1. Characteristics of CW and SLF used in the test

Parameter	SLF	CW
TS (g L^{-1})	36.1	55.7
VS (g L^{-1})	23.2	46.4
COD (g L^{-1})	30.4	58.0
TKN-N (g L^{-1})	2.3	0.65
NH_4^+- N (g L^{-1})	1.6	0.15
P_T (g L^{-1})	0.57	0.18
Alkalinity (mg $CaCO_3$ L^{-1})	12,500	0
pH	6.9	5.3

Dairy manure was collected from an intensive dairy farm located in Loredo (Cantabria, Spain, Lat: N 43°27′35′′ Long: W 3°42′20′′) equipped with a 0.8 mm mesh size screw press separator (Cri-Man, Italy). It holds 140 lactating cows, 40 dry cows, 80 heifers and 30 calves.

The screened liquid fraction of dairy manure (SLF) was used in this test to facilitate operation and to enhance the contact between microorganisms and substrate. Characteristics of the SLF are also shown in Table 1.

2.2. Inoculum

Anaerobically digested mesophilic biomass from a pilot plant that processes dairy manure (IES la Granja, Heras, Cantabria, Spain, Lat: N 43º24′16′′ Long: W 3º46′11′′) was used as inoculum for this test. It had 52.2 g VS L^{-1} and a VS/TS ratio of 0.66.

2.3. Batch Laboratory Anaerobic Co-Digestion Tests

250-mL serum bottles capped with rubber septum sleeve stoppers were used as reactors. Gas production was determined by pressure measurement. The pressure was taken from the headspace of the reactors through the septum with a syringe connected to a digital pressure sensor with silicon measuring cell (ifm, Germany - type PN78, up to 2 bar). Biogas samples were also taken through the septum by a needle connected to a syringe. All the reactors were manually stirred once a day.

After set-up of the reactors helium was flushed to remove the air in the headspace of the bottles. Thereafter, all the reactors were placed in a thermostatic bath at 35ºC. Gas produced in each reactor was determined daily for 30 days.

In order to compare the different proportions of CW in the anaerobic co-digestion test, mixtures of CW and SLF were prepared with the following mass proportions of whey: 10% (M10), 20% (M20), 30% (M30), 40% (M40), 50% (M50) and 60% (M60). Additionally the SLF and the CW were tested separately to determine their ultimate methane yield (B_0).

Table 2. Amounts of substrates and inoculum used in the test

Sample	Inoculum (g)	Cheese Whey (g)	SLF (g)	Substrate COD/TKN-N ratio	BM (g)	$VS_{inoculum}$/ $VS_{substrate}$
SLF	42.4	0	83.4	13.2	0	1.14
M10	43.2	8.44	74.9	15.6	0	1.06
M20	42.6	17.1	66.8	18.3	0	0.95
M30	42.5	24.9	58.7	21.4	0	0.88
M40	43.2	33.6	50.0	25.3	0	0.83
M50	43.2	41.8	42.2	29.8	0	0.77
M60	43.0	50.4	34.2	35.6	0	0.72
CW_{BM}	50.9	57.1	0	89.2	25	1.00
CW_{nBM}	80.5	44.8	0	89.2	0	2.02

To determine the B_0 of the CW, a sample of CW (CW_{BM}) was provided with a basal medium (BM) containing nutrient, trace metal, and alkalinity to avoid inhibition of the process. The basal medium composition was the same than that reported by Ergüder et al. [14]. Another sample of CW (CW_{nBM}) was processed without pH control to check inhibition of the process. Batch anaerobic tests were performed in duplicate rather than in triplicate since CW and SLF were quite homogeneous samples. Two blanks with water and inoculum were also tested to measure methane potential of inoculum. Results are expressed as means subtracting methane production from the blanks.

For each mixture of CW and SLF, reactor was supplied with approximately 43 g of inoculum and 84 g of substrate mixture. Table 2 shows the amounts of inoculum, CW and SLF used for each mixture digestion. The resulting $VS_{inoculum}/VS_{substrate}$ ratios ranged from 1.07 for M10 to 0.75 for M60. For SLF, $VS_{inoculum}/VS_{substrate}$ ratio was 1.14. For CW_{BM} test a ratio $VS_{inoculum}/VS_{substrate}$ of 1 was used whereas for CW_{nBM} a higher ratio was used. Since experiments were performed in batch, only methane production was used as indicator of the co-digestion process.

2.4. Analytical Techniques

Biogas composition was measured on a 2m Poropak T column in a HP 6890 GC System with helium as the carrier gas and a TCD detector. Methane volumes are expressed at 0°C and 1 atm. All other analyses (TS, VS, COD, Total Kjeldahl Nitrogen (TKN-N), Ammonia Nitrogen (NH_4^+-N), and Total Phosphorous (P_T) were performed according to Standard Methods [33].

3. Results and Discussion

3.1. Characteristics of Substrates

Characteristics of CW and SLF are presented in Table 1. CW presents higher organic content than SLF but a lower nutrient and alkalinity content. As was mentioned in the introduction, the characteristics of CW depend on several factors and for this reason there can be found different CW compositions in the literature in terms of VS, COD and nutrient content.

In this work CW presented VS and TS content of 46.4 and 55.7 g L^{-1}. In all cases, a common characteristic in cheese whey is its negligible or low bicarbonate alkalinity that would make its anaerobic digestion not feasible without pH control as reported by Ergüder et al. [14], Ghaly [16] and Malaspina et al. [17]. With regards to the SLF it presented lower organic content than other screened liquid fraction used by the authors [34], but enough alkalinity to enable the anaerobic process. This low VS content will limit methane production per kg of substrate.

3.2. Batch Tests

3.2.1. Methane Potentials of CW and SLF

The methane production curves for SLF and CW are represented in Figure 1. With regards to CW, a strong inhibition in the sample CW_{nBM} was observed. Compared with CW_{BM}, the whey sample that received the BM containing alkalinity, it is clear that anaerobic digestion of cheese whey without pH control is not feasible, as was reported by Ergüder et al. [14] and Ghaly [16]. In Figure 1a, the daily specific methane production for both CW_{BM} and CW_{nBM} samples can be observed. A similar trend in methane production during the first three days of the experiment was noted. A peak at day 1 followed by a decrease in methane production at days 2 and 3. However at day 4, the CW that received the BM recovered methane production whereas methane production from CW_{nBM} dropped to zero at day 3 and did not recover. In the case of CW_{BM}, alkalinity avoided the drop of pH and after a lag phase of two days, methane production started to increase reaching a new peak at day 6; then the methane yield started to decrease and dropped to zero after 26 days from the trial start.

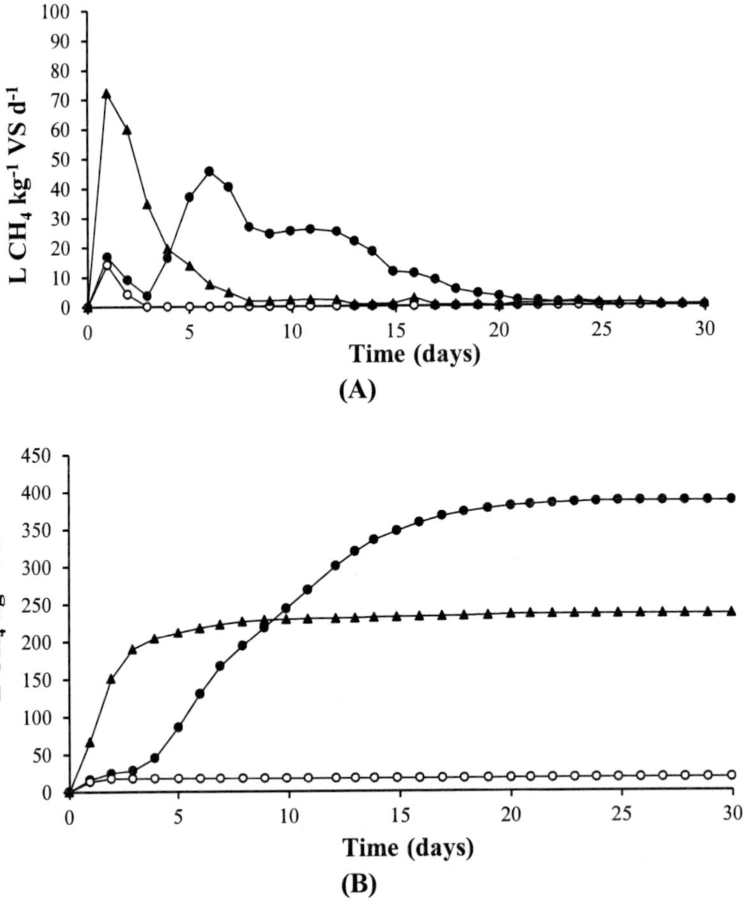

Figure 1. Methane yields in batch tests for cheese whey and screened liquid fraction of dairy manure at 35ºC. A. Daily specific methane productions. B. Cumulative specific methane yields. (●: CW_{BM}, ○: CW_{nBM}, ▲: SLF).

The lag phase during days 2 and 3 can be due to inoculum acclimatization to the substrate. The sample of CW that did not receive alkalinity supplementation yielded a very low methane production. After the 30 days test, the pH of CW_{nBM} sample was measured. It had drooped down to 3.2, so the inhibition of the process was caused by acidification and subsequent low pH. In Figure 1b, where cumulative methane productions are shown, it can be observed that CW_{BM} yielded 385 L CH_4 kg^{-1} VS, whereas CW_{nBM} only yielded 17 L CH_4 kg^{-1} VS. Taking into account the VS content of the CW, the B_0 of the CW employed in this test was 17.9 L CH_4 kg^{-1} CW. In this sense, Dinuccio et al. [35] and Ergüder et al. [14] reported methane yields of 31.0 and 23.4 L CH_4 kg^{-1} whey respectively.

3.2.2. Co-Digestion Tests

In Figure 2 the daily specific methane production rates for all the co-digestion samples are presented. They have been separated in two groups, M10, M20 and M30 in Figure 2a and M40, M50 and M60 in Fig 2b. Daily specific methane yield showed a similar trend for samples M10, M20 and M30. This trend was characterized by a peak at day 1 and then by a progressive methane yield decrease which dropped to negligible values after 12 days from the start of the test. Moreover the higher the CW content, the higher the peak observed in methane yield. These peaks were 80, 95 and 107 L CH_4 kg^{-1} VS d^{-1} for M10, M20 and M30 respectively. The higher CW proportion, the higher peak was observed.

The daily specific methane production rates for the co-digestion samples M40, M50, M60 are shown in Figure 2b. These yields are represented separately due to the different behavior between both groups of samples. In samples M40 and M50 the peak was reached at day 2 and for the sample M60 three peaks in methane production can be observed. In this case as CW proportion increased, the peak in methane production decreased. These peaks were 62 and 48 L CH_4 kg^{-1} VS d^{-1} for M40 and M50 respectively. After these peaks, there was a progressive methane yield decrease which dropped to negligible values after the 15 days of test. For the sample M60, the sample with the highest CW proportion, three peaks were observed, one of 35 L CH_4 kg^{-1} VS d^{-1} at day 1, another one of 34 L CH_4 kg^{-1} VS d^{-1} at day 7 and the last one of 25 L CH_4 kg^{-1} VS d^{-1} at day 10. All these peaks in methane production can be caused by the sudden degradation of accumulated volatile fatty acids. In this case, the higher the CW proportion, the lower peak obtained. This can be explained by the fact that batch co-digestion tests were performed on a fixed mass ratio of inoculum to substrate, resulting in lower ratios $VS_{inoculum}/VS_{substrate}$ for the samples with higher CW proportion. However, this tendency did not happen with samples M10, M20, M30 where the higher CW proportion, the higher peak was obtained. So, it can be a sign indicating the first effects of inhibition at CW proportions higher than 30%. It must be noted that the co-digestion trial was not a biochemical methane production (BMP) test. In order to avoid inhibitions during BMP tests, buffer solutions and larger amounts of inoculum are added as reported by Lesteur et al. [36]. In this work a constant mass ratio of inoculum to substrate has been used to study the influence of CW proportion in the anaerobic co-digestion of CW and dairy manure.

Figure 3 shows the cumulative specific methane yield for SLF and the co-digestion samples. SLF obtained the lowest methane yield value, 236 L CH_4 kg^{-1} VS whereas the sample containing 20% of CW yielded the highest value, 295 L CH_4 kg^{-1} VS. In the middle, the different mixtures of CW and screened liquid manure. Methane yields for the mixtures whey – liquid manure varied with whey proportion. Due to higher methane potential of CW, mixtures with upper whey content should yield larger methane yields.

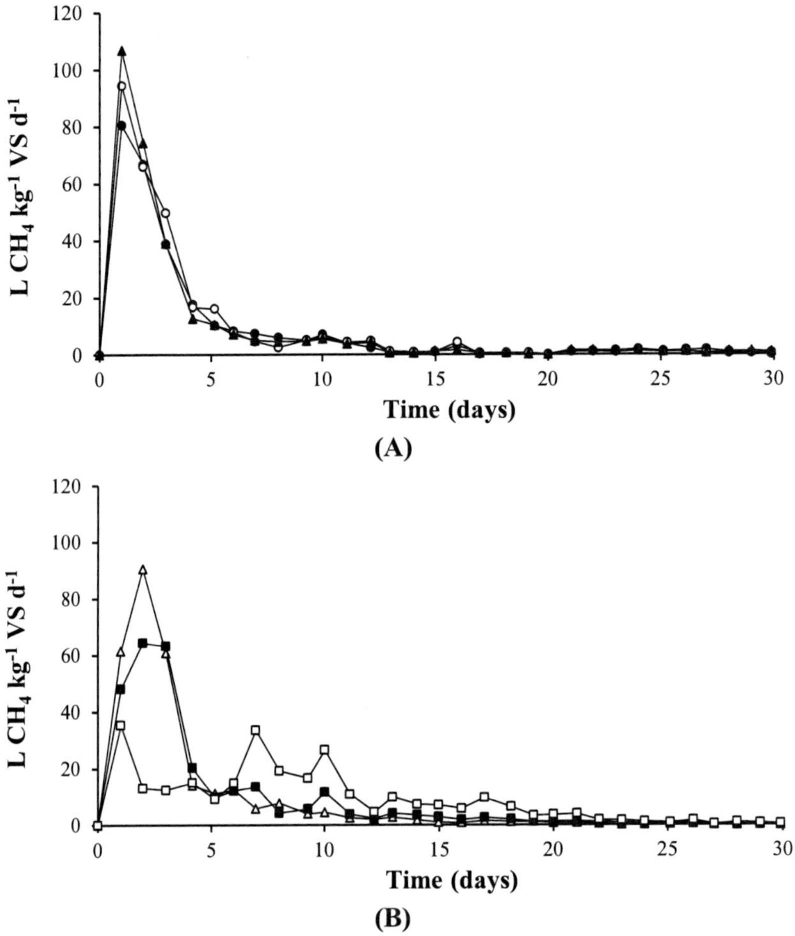

Figure 2. Daily specific methane yields for co-digestion samples at 35ºC. A. M10, M20 and M30 (●: M10, ○: M20, ▲: M30). B. M40, M50 and M60. (△: M40, ■: M50, □: M60)

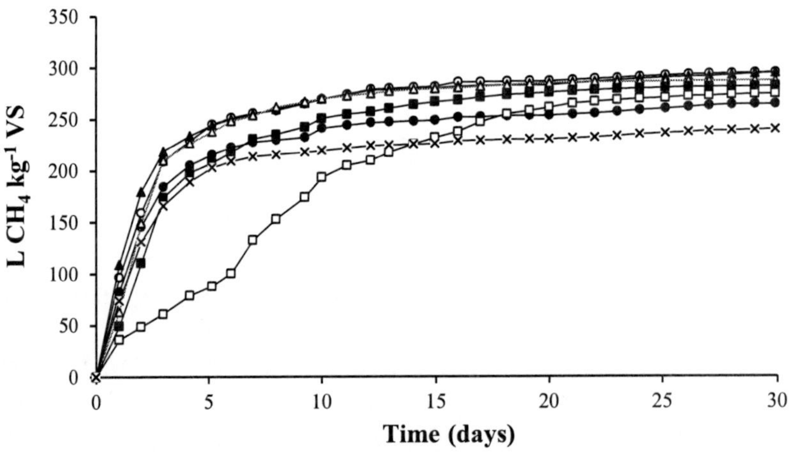

Figure 3. Cumulative specific methane productions for co-digestion samples and SLF at 35ºC (●: M10, ○: M20, ▲: M30, △: M40, ■: M50, □: M60, ×: SLF)

However specific methane yield only increased for the samples with 10% and 20% CW proportions. For the samples M30 to M60, methane yield decreased compared to M20, reaching the following values: 294, 287, 281 and 274 L CH_4 kg^{-1} VS for M30, M40, M50 and M60 respectively. In addition, in Figure 3 it can be observed the different behavior between M60 and the rest samples. All the samples except M60 showed a rapid and constant increase in methane production during the first five days of the test, whereas methane production for sample M60 was slower, as could be deducted from data in Figure 2B.

Cumulative methane yields per mass of sample for all the co-digestion and SLF samples are presented in Figure 4. After the 30 days test methane production of the different samples were: 5.5, 6.8, 8.1, 8.9, 9.3, 9.8 and 10.1 L CH_4 kg^{-1} sample for SLF, M10, M20, M30, M40, M40, M50 and M60 respectively. As can be observed the increase in CW proportion resulted in higher methane yields. However the linearity was maintained only up to 20% CW content as Figure 5 presents.

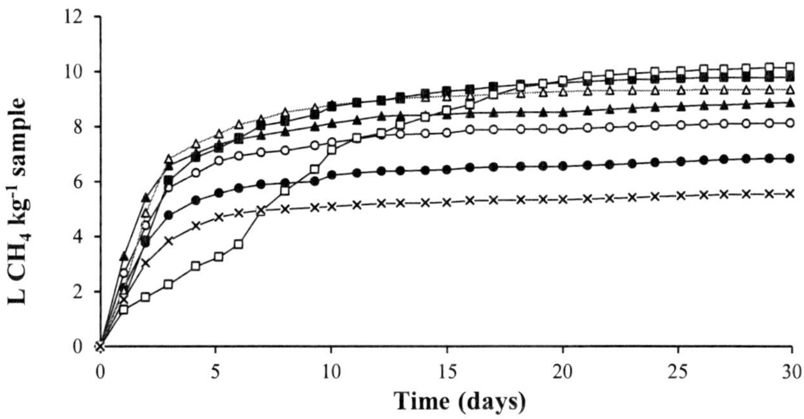

Figure 4. Cumulative methane yields for co-digestion samples and SLF at 35°C (●: M10, ○: M20, ▲: M30, △: M40, ■: M50, □: M60, ×: SLF)

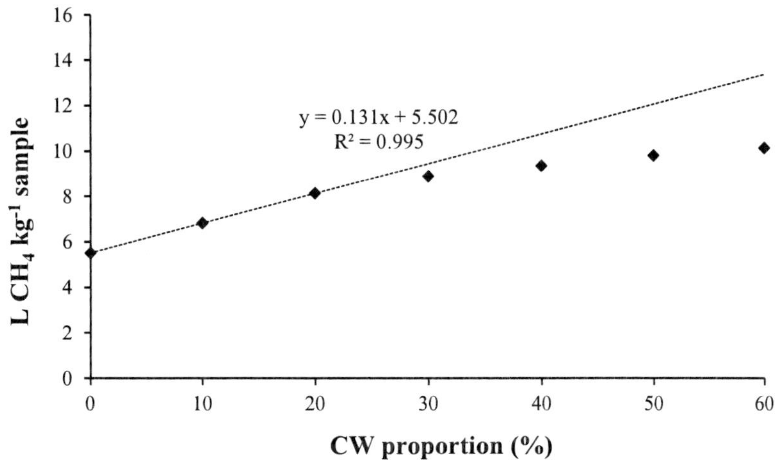

Figure 5. Ultimate methane yield for co-digestion samples vs CW proportion.

Methane production of CW can be determined for each co-digestion sample if we use out the B_0 of SLF as a blank with the help of the equations 1 and 2: where Y is the methane yield of each co-digested sample; $\%_{CW}$ and $\%_{SLF}$ are the content percentage of CW and SLF in each sample and Y_{CW} is the methane yielded to calculate for the CW in each sample. In Table 3 the values of Y_{CW} calculated for each co-digestion sample are detailed.

Table 3. Methane yield calculated from CW in each sample

Sample	Y_{CW} (L CH_4 kg^{-1} CW)
M10	18.2
M20	18.4
M30	16.6
M40	15.0
M50	14.0
M60	13.2

$$Y = \frac{Y_{CW} \cdot \%_{CW} + 5.5 \cdot \%_{SLF}}{100} \quad (1)$$

$$Y_{CW} = \frac{Y \cdot 100 - 5.5 \cdot \%_{SLF}}{\%_{CW}} \quad (2)$$

Table 3 illustrates the decrease in the efficiency of the process as CW proportion increases. As expected from previous data obtained, samples with 10% and 20% of CW gave similar methane yield per kg of CW which would result in higher methane production per mass of mixture as CW proportion increases within this range. When CW content was equal or higher than 30%, the methane yield per kg of CW suffered a decrease. The higher the CW content the lower the methane yield from CW. Alkalinity from SLF was enough to maintain the expected methane production up to 20% CW proportion. At higher CW proportions the mixture was degraded to methane but a loss of efficiency was detected. Indeed, at a CW proportion of 60% a slower rate of methane production was observed and the ultimate methane yield of the co-digested sample M60 was similar to that of M50. The best results in terms of methane conversion were obtained at 20% cheese whey proportion. Under this conditions methane yield of the mixture was 46.4% higher than that of the liquid manure, and cheese whey was completely degraded with an estimate methane yield of 18.4 L CH_4 kg^{-1} whey. On the contrary, with 60% cheese whey proportion the methane yield of the mixture was 82.7% higher than that of the liquid manure but the estimate methane yield of the whey was reduced to 13.2 L CH_4 kg^{-1} whey. In this sense Gelegenis et al. [29] reported a dramatic decrease in biogas production during the co-digestion of CW and poultry manure at proportions higher than 50%. On the contrary, Comino et al. [30] stated that anaerobic co-digestion of cattle slurry and CW is possible with up to 65% CW content. Moreover, Kavacik and Topaloglu [6] stated that it is necessary to start the co-digestion process with a mixture a CW and digested manure. It must be taken into account that the inoculum used in this work was digested manure which also provided alkalinity to the process. Another parameter to analyze is the ratio COD/TKN-N in the substrate mixture which can be found in Table 2. From these data it can be deducted that extreme COD/TKN-N ratios (13.2 for SLF and 89.2 for CW) were not limiting for the anaerobic process when there was alkalinity enough to maintain the pH within the anaerobic pH range. Severe inhibition by acidification was detected for the CW sample that did not receive alkalinity supplementation. For the co-digestion samples, over 20% CW proportion, the higher CW proportion the lower methane yield obtained with respect to the expected yield. There was not a severe inhibition but methane yields lower than expected were probably due to some inhibition caused by lack of alkalinity in the substrate mixture.

Further experiments will be performed to search the limit of cheese whey proportion in the anaerobic co-digestion of cheese whey and the screened liquid fraction of dairy manure in continuous operation mode.

CONCLUSION

This study shows that the screened liquid fraction of dairy manure acts as an stabilizer, due to its alkalinity content, for the anaerobic co-digestion with cheese whey, allowing to digest a difficult substrate such as the whey without alkalinity addition. However, cheese whey proportions higher than 20% resulted in methane yields lower than those expected from methane potentials of both substrates separately, which indicates some inhibition. So, caution must be taken when co-digesting this kind of substrate without chemicals addition for pH control. Further experiments must be carried out to obtain the maximum cheese whey content in the co-digestion process in continuous operation mode.

ACKNOWLEDGMENTS

The authors would like to thank the Ministry of Science and Innovation for financially supporting this research under contract CTM2011-28042. Co-funded with FEDER funds.

REFERENCES

[1] Lesschen, J.P., van den Berg, M., Westhoek, H.J., Witzke, H.P., Oenema, O., Greenhouse gas emission profiles of European livestock sectors. *Anim. Feed Sci. Tech.* 2011, *166-167*, 16-28.

[2] Rabaud, N.E., Ebeler, S.E., Ashbaugh, L.L., Flocchini, R.G., Characterization and quantification of odorous and non-odorous volatile organic compounds near a commercial dairy in California. *Atmos. Environ.* 2003, *37*, 933–940.

[3] Ball Coelho, B., Murray, R., Lapenc, D., Topp, E., Bruin, A., Phosphorus and sediment loading to surface waters from liquid swine manure application under different drainage and tillage practices. *Agr. Water Manage.* 2012, *104*, 51-61.

[4] De Vries, J.W., Groenestein, C.M., De Boer, I.J.M., Environmental consequences of processing manure to produce mineral fertilizer and bio-energy. *J. Environ. Manage.* 2012, *102*, 173-183.

[5] Peters, D., Raw materials. *Adv. Biochem. Eng. Biotechnol.* 2006, *105*, 1–30.

[6] Kavacik, B., Topaloglu, B., Biogas production from co-digestion of a mixture of cheese whey and dairy manure. *Biomass Bioenerg.* 2010, *34*, 1321-1329.

[7] Koutinas, A., Papapostolou, H., Dimitrellou, D., Kopsahelis, N., Katechaki, E., Bekatorou, A., Bosnea, L., Whey valorisation: a complete and novel technology development for dairy industry starter culture production. *Bioresour. Technol.* 2009, *100*, 3734–3739.

[8] Mawson, A.J., Bioconversions for whey utilization and waste abatement. *Bioresour. Technol.* 1994, *47*, 195-203.

[9] Kosaric, N., Asher, Y. J., The utilization of cheese whey and its components. *Convers. Recycl.* 1982, *5*, 23-32.

[10] Sandhu, D., Waraich, M., Conversion of cheese whey to single cell protein. *Biotechnol. Bioeng.* 1983, *25*, 797-808.

[11] Siso, M.I.G., The biotechnological utilization of cheese whey: a review. *Bioresour. Technol.* 1996, *57*, 1-11.

[12] Prazeres, A.R., Carvalho, F., Rivas, J., Cheese whey management: A review. *J. Environ. Manage.* 2012, *110*, 48-68.

[13] Saddoud, A., Hassairi, I., Sayadi, S., Anaerobic membrane reactor with phase separation for the treatment of cheese whey. *Bioresour. Technol.* 2007, *98*, 2102-2108.

[14] Ergüder, T., Tezel, U., Güven, E., Demirer, G.N., Anaerobic biotransformation and methane generation potential of cheese whey in batch and UASB reactors. *Waste Manage.* 2001, *21*, 643-650.

[15] García, P.A., Rico, J.L., Fdz-Polanco, F., Anaerobic treatment of cheese whey in a two-phase UASB reactor. *Environ. Technol.* 1991, *12*, 355-362.

[16] Ghaly, A.E., A comparative study of anaerobic digestion of acid cheese whey and dairy manure in a two-stage reactor. *Bioresour. Technol.* 1996, *58*, 61–72.

[17] Malaspina, F., Cellamare, C.M., Stante, L., Tilche, A., Anaerobic treatment of cheese whey with a downflow-upflow hybrid reactor. *Bioresour. Technol.* 1996, *55*, 131-139.

[18] Yan, J.Q., Liao, P.H., Lo, K.V., Methane production from cheese whey. *Biomass* 1988, *17*, 185-202.

[19] Malaspina, F., Stante, L., Cellamare, C.M., Tilche, A., Cheese whey and cheese factory wastewater treatment with a biological anaerobic-aerobic process. *Water Sci. Technol.* 1995, *32*, 59-72.

[20] Kelly, C.R., Switzenbaum, M.S., Anaerobic treatment: temperature and nutrients effects. *Agr. Wastes* 1984, *10*, 135–154.

[21] Bachmaier, J., Effenberger, M., Gronauer, A., Greenhouse gas balance and resource demand of biogas plants in agricultures. *Eng. Life Sci.*, 2010, *10*, 560-569.

[22] Garfí, M., Gelman, P., Comas, J., Carrasco, W., Ferrer, I., Agricultural reuse of the digestate from low-cost tubular digesters in rural Andean communities. *Waste Manage.* 2011, *31*, 2584-2589.

[23] Guo, J., Clemens, J., Li, X., Xu, P., Dong, R., Performance evaluation of a Chinese medium-size agricultural biogas plant at ambient temperature. *Eng. Life Sci.* 2012, *12*, 336-342.

[24] Lansche, M. and Müller, J., Life cycle assessment of energy generation of biogas fed combined heat and power plants: Environmental impact of different agricultural substrates. *Eng. Life Sci.* 2012, *12*, 313-320.

[25] Abbasi, T., Tauseef, S.M., Abbasi, S.A., Anaerobic digestion for global warming control and energy generation. An overview. *Renew. Sust. Energ. Rev.* 2012, *16*, 3228-3242.

[26] Nasir, I.M., Mohd Ghazi, T.I., Omar, R., Anaerobic digestion technology in livestock manure treatment for biogas production: A review. *Eng. Life Sci.* 2012, *12*, 258-269.

[27] El-Mashad, H.M., Zhang, R., Biogas production from co-digestion of dairy manure and food waste. *Bioresour. Technol.* 2010, *101*, 4021-4028.

[28] Marañón, E., Castrillón, L., Quiroga, G., Fernández-Nava, Y., Gómez, L., García, M.M., Co-digestion of cattle manure with food waste and sludge to increase biogas production. *Waste Manage.* 2012, *32*, 1821-1825.

[29] Gelegenis, J., Georgakakis, D., Angelidak, I., Marris, V., Optimization of biogas production by co-digesting whey with diluted poultry manure. *Renew. Energ.* 2007, *32*, 2147-2160.

[30] Comino, E., Riggio, V., Rosso, M., Biogas production by anaerobic co-digestion of cattle slurry and cheese whey. *Bioresour. Technol.* 2012, *114*, 46-53.

[31] Hublin, A., Zelic, B., Modelling of the whey and cow manure co-digestion process. *Waste Manage. Res.* 2012, in press DOI: 10.1177/0734242X12455088.

[32] Lo, K.V., Liao, P.H., Chiu, C., Mesophilic anaerobic digestion of a mixture of cheese whey and dairy manure. *Biomass* 1988, *15*, 45-53.

[33] APHA, *Standard methods for the examination of water and wastewater*, 20th ed., American Public Health Association, Washington, DC, 1998.

[34] Rico, C., Rico, J.L., Muñoz, N., Gómez, B., Tejero, I., Effect of mixing on biogas production during mesophilic anaerobic digestion of screened dairy manure in a pilot plant. *Eng. Life Sci.* 2011, *11*, 476–481.

[35] Dinuccio, E., Balsari, P., Gioelli, F., Menardo, S., Evaluation of the biogas productivity potential of some Italian agro-industrial biomasses. *Bioresour. Technol.* 2010, *101*, 3780-3783.

[36] Lesteur, M., Bellon-Maurel, V., Gonzalez, C., Latrille, Roger, E.J.M., Junqua, G., Steyer, J.P., Alternative methods for determining anaerobic biodegradability: A review. *Process Biochem.* 2010, *45*, 431-440.

INDEX

C

D

E

H

I

J

K

L

P

Q

R

S

T